A. Muttoni
J. Schwartz
B. Thürlimann

Design of Concrete Structures with Stress Fields

Birkhäuser
Basel Boston Berlin

Authors:

Dr. A. Muttoni
Grignoli & Muttoni
via Somaini 9
6900 Lugano
Schweiz

Dr. J. Schwartz
Ing.-bureau Frey & Schwartz
Steinhauserstr. 25
6300 Zug
Schweiz

Prof. Dr. B. Thürlimann
Pfannenstielstr. 56
8132 Egg
Schweiz

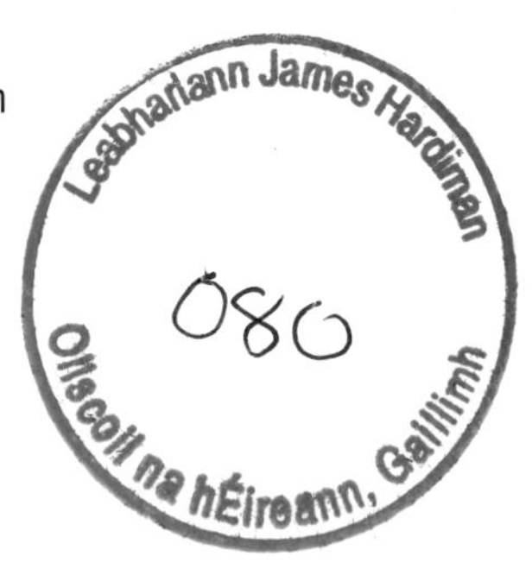

Die Deutsche Bibliothek – Cataloging-in-Publication Data

Muttoni, Aurelio:
Design of Concrete Structures with Stress Fields/
A. Muttoni ; J. Schwartz ; B. Thürlimann.
- Basel ; Boston ; Berlin : Birkhäuser, 1996
Engl. Ausg. u. d. T.: Muttoni, Aurelio: Design of concrete
structures with stress fields
ISBN 3-7643-5491-7 (Basel ...)
ISBN 0-8176-5491-7 (Boston)
NE: Schwartz, Joseph:;Thürlimann, Bruno:

Printed on acid-free paper
produced from chlorine-free pulp. TCF ∞
Cover design: Karin Weisener
Printed in Germany
ISBN 3-7643-5491-7
ISBN 0-8176-5491-7

9 8 7 6 5 4 3 2 1

CONTENTS

PREFACE AND ACKNOWLEDGEMENTS

This book evolved from "Lecture Notes" (Ref. [36]) prepared for a semester course offered to senior and graduate students at the Swiss Federal Institute of Technology, Zürich, in 1987. The notes were also used for a series of conferences at other institutions and in particular for seminars in München (VBI-Seminar, March 16, 1989), in New York (ASCE-Metropolitan Section, Oct. 10-11, 1989) and in Oslo (Norwegian Contractors, June 10-12, 1992). The authors gained much valuable input from discussions and personal contacts at all these events. In addition, they had the opportunity to apply the design procedures presented in this book to practical cases covering the entire range of reinforced concrete structures from foundations, buildings, bridges to offshore-structures. Hence, the material presented in the book has matured over an extended period of time with respect to content, presentation and practical applicability.

The authors express their sincere appreciation to the Research Foundation of the Swiss Cement Producers for their encouraging support and generous financial contributions over the extended period of this book's development. They furthermore gratefully acknowledge the help of Dr. G. Prater in translating the German version of the manuscript into English and the preparation of all drawings by Mr. S. Schorno.

Mrs. Carolyn Gorczyca, P.E., made a review of the manuscript. The authors are most thankful for her valuable suggestions and improvements.

They would also like to recognize the cooperative and diligent support by the publisher, Birkhäuser Verlag.

Finally, they hope that the reader will gain a clearer insight into the behavior and strength of reinforced and prestressed concrete structures, founded on a unified theoretical basis. This in turn may help him to conceive and design such reinforced concrete structures with greater confidence, self-assurance and freedom.

1 INTRODUCTION AND THEORETICAL BASIS

1.1 INTRODUCTORY REMARKS

"There is nothing more practical than a simple theory"

Robert Maillart

The aim of statical analysis is to design, calculate and erect a structure in such a way that the demands of strength, functionality and durability are fulfilled. Important parts of the design process are the dimensioning, i.e. fixing the dimensions and the reinforcement, and the detailing, i.e. the arrangement of the reinforcement, and the layout of reinforcing bars, joints and supports etc. Up to the present time empirical rules based on years of practical experience and test results have been applied for design and construction purposes. What these lack, however, is a unified theoretical basis.

The present situation is aptly described by J.G. MacGregor in an article entitled "Challenges and Changes in the Design of Concrete Structures" [1]:

> *"Material behavior can be idealized as consisting of an 'elastic' domain and a 'plastic' domain. For almost 200 years structural design has been based on elastic theory which assumes that structures display a linear response throughout their loading history, ignoring the post-yielding stage of behavior. Current design practice for reinforced concrete structures is a curious blend of elastic analysis to compute forces and moments, plasticity theory to proportion cross-sections for moment and axial load, and empirical mumbo-jumbo to proportion members for shear. One of the most important advances in reinforced concrete design in the next decade will be the extension of plasticity based design procedures to shear, torsion, bearing stresses, and the design of structural discontinuities such as joints and corners. These will have the advantage of allowing a designer to follow the forces through a structure.*
>
> *The opposite of plasticity theory is finite element analysis. There, by breaking the structure down into small elements, it is possible to get an estimate of internal stresses and strains. The estimate is only as good as the assumptions which, for economic and practical reasons, generally oversimplify the problem to reduce the number of elements or to simplify element layout. Occasionally these simplifications will be such that*

the nature of the problem is completely distorted. The finite element approach is basically incompatible with modern limit state methods of proportioning concrete structures. The analysis gives elastic stresses and strains, the design methods are based on forces and moments. If the analysis ignores cracking, the results will be almost meaningless in those problems where conventional beam theory cannot be applied. It is essential that proper guidance be developed to aid engineers in making the transition from finite element analyses to the selection of reinforcement."

Already in 1938 Ernst Melan [2] made a remarkable observation, which until today has gone practically unnoticed in design practice, namely:

"....that the problem of determining the stresses in an elasto-plastic body requires the knowledge of the existing permanent deformations at that particular instant. Since these, however, depend on the previous load cases, it is necessary to know the past load history. In practice, in the majority of cases, the conditions are such that different load cases are possible, which have been applied many times in an arbitrary sequence. Thus the state of stress caused by a particular load case is in general not the same as when this loading is repeated after a series of load changes. Since, as already stated, the load history is characterized by a certain arbitrariness, the question of obtaining the state of stress under a particular loading is really rather meaningless."

In each structure unknown initial residual stresses are present. Even if the stresses in a structure could be determined exactly for a given loading, they are inevitably superimposed on such undefined initial stresses. Therefore, it follows that structural design cannot be based alone on a detailed stress analysis. Robert Maillart (1872-1940) expressed his views on the subject as follows [3]:

"Is it really necessary to consider all the secondary circumstances in an analysis? To be sure, the view often prevails that the calculation clearly and conclusively governs the design. However, when one realizes the impossibility of taking into account all these secondary influences, it becomes evident that each calculation is only an aid to the designer, who also has to take into account these secondary effects in some way. Depending on the circumstances, the results of the calculation may have direct application or be modified. The second is often the case, especially if it is a designer and not an analyst who is doing the work. The former will not allow a thin wall to be stressed to the same extent as a slab, but will proportion it to be a little thicker. For the slab, however, he will allow somewhat higher stresses than furnished by the calculation.

> *He will also consider, especially in the case of beams whose span is small compared to the height, whether he should make full use of the bond resistance permitted him by the calculations. Unfortunately, the standards and codes of practice, especially when they are used in teaching courses or applied literally by state control engineers, mislead or even force the design engineer to adhere to them in a mechanical fashion. A general relaxing of the regulations in favor of placing more responsibility on the design engineer would lead to a significant qualitative improvement of structures. Above all, the student should not be confronted with codes and standards during his university studies, as this can only be detrimental to a free unbiased outlook."*

Emil Mörsch (1872-1950) also recognized that in the design and proportioning of structures [4]:

> *"Only an exact knowledge of the structural materials and their behavior up to failure in the individual structural members enables the engineer involved with reinforced concrete design to adapt the conventional calculation methods correctly for the particular problem, and accept the responsibility of designing and executing complex reinforced concrete structures, which fulfil the required safety regulations in all parts without being uneconomic."*

It will now be shown that the theory of plasticity provides a consistent scientific basis, from which simple and, above all, clear models may be derived to determine the statical strength of reinforced and prestressed concrete structures. These models, of course, may be used for design and, thanks to their clarity, for detailing the structural components. Whether they produce useful results can only be demonstrated by systematic testing and, above all, by practical experience.

Based on many years of experience, it has been confirmed that well designed structures can sustain unexpected load cases, e.g. explosions, earthquakes, settlements etc., with more or less moderate damage, although according to elastic theory they should have collapsed. The plastic redistribution of internal forces that is thereby necessary is often aptly described as "self-healing" or somewhat ironically as the "smartness" of the material.

In the last 40 years extensive tests have been carried out on materials, components and structures, firstly in structural steel and subsequently also in reinforced concrete. Especially important for design practice are the tests on reinforced concrete, as failure is very often governed by the performance of details, splices and connections. All these tests have demonstrated that the theory of plasticity provides a rational, logical, consistent and simple method for investigating the statical behavior and strength of steel and reinforced concrete structures.

In the following, the fundamentals of the theory of plasticity are briefly given and the limit state theorems are derived in an intuitive physical manner. The subsequently presented plastic design models are based on these theorems. A more exhaustive treatment of the theoretical basis is given in [5].

1.2 FUNDAMENTALS OF THE THEORY OF PLASTICITY

1.2.1 Material Behavior

In the theory of plasticity the mechanical behavior is restricted to those quantities, which for given assumptions essentially determine the strength. If for a particular material the plastic deformations are substantially greater than the elastic deformations, then the latter may be neglected.

Fig. 1.1 shows the corresponding uniaxial stress-strain diagram for structural or reinforcing steel. The assumption of rigid-plastic behavior is always a drastic simplification of reality, and can only be justified if total strains are obtained which are considerably higher than the elastic limit strain ε_{sy}. Then it is only necessary to specify the strain rate $\dot{\varepsilon}_s$, i.e. the direction of the strain increment.

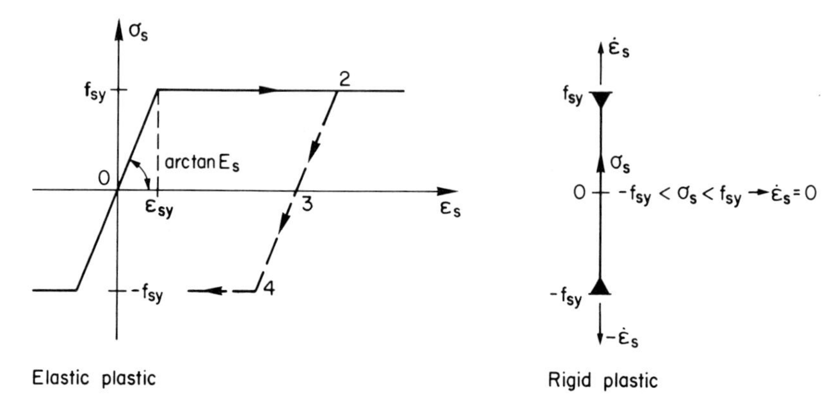

Fig. 1.1: Idealized stress-strain diagram for steel

The assumed behavior for concrete shown in Fig. 1.2 represents a much greater simplification of reality than is the case for steel.

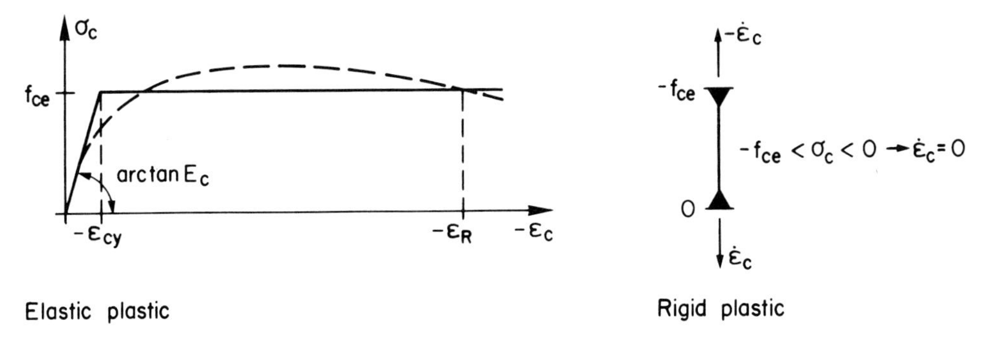

Fig. 1.2: Idealized stress-strain diagram for concrete

The determination of the actual strength values, i.e. the yield stress of the reinforcement f_{sy} and the effective concrete strength f_{ce}, is discussed in detail in chapter 3.

Moment-Curvature Relationship

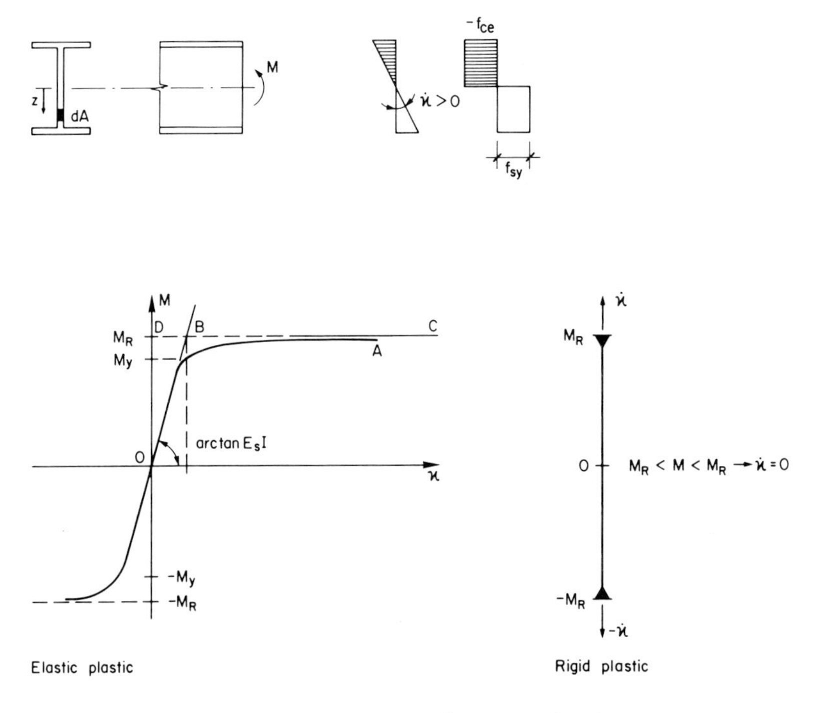

Fig. 1.3: Moment-curvature diagram for a steel I-beam

The statical behavior of plane frame and beam structures is characterized above all by the moment-curvature relationship. This is shown in Fig. 1.3 for a steel girder.

For the calculation of the deflection curve either the exact M-$\varkappa$ diagram O–A or the bilinear approximation O-B-C, which corresponds to the M-$\varkappa$ curve for an I-section without web, can be used.

The rigid-plastic solution utilizes only the plastic moment M_R and the plastic curvature increment $\dot{\varkappa}$. The corresponding moment-curvature relationship for a reinforced concrete beam is shown in Fig. 1.4.

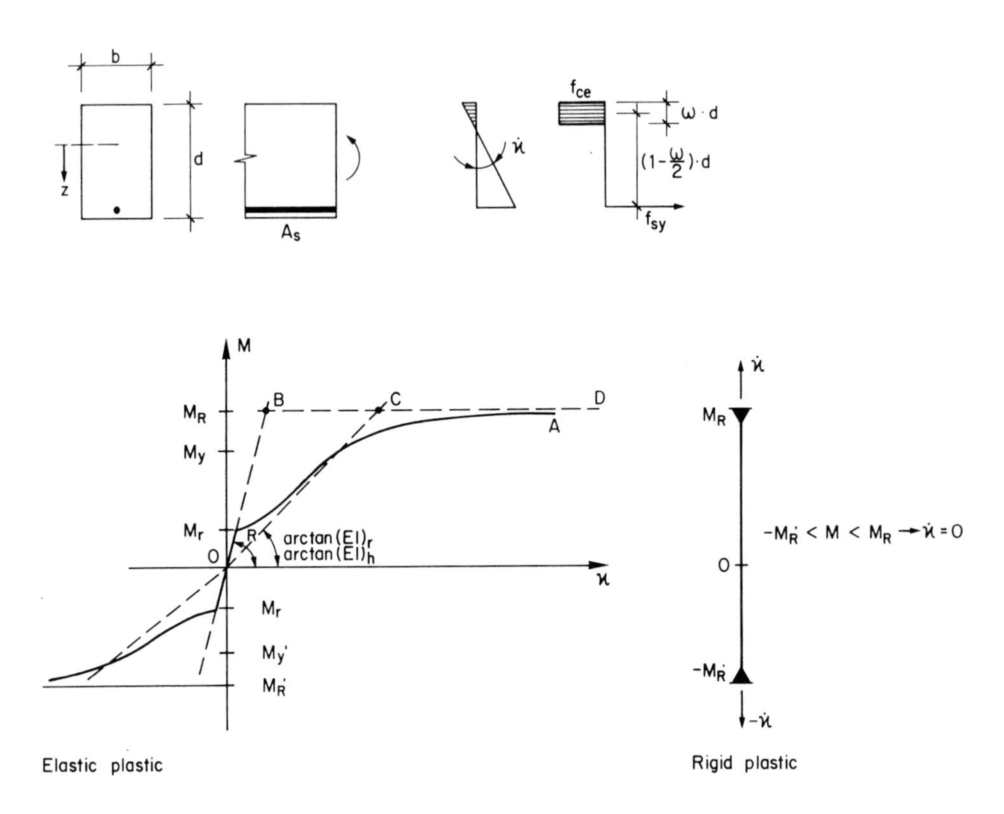

Fig. 1.4: Moment-curvature diagram for a reinforced concrete beam

Up to the point where the cracking moment M_r is reached the stiffness $(EI)_h$ of the homogeneous concrete cross-section is effective. Afterwards it decreases with increasing crack development to the stiffness of the reinforced concrete section without tensile strength $(EI)_r$. After reaching the yield moment M_y, i.e. $\sigma_s = f_{sy}$, the stiffness decreases sharply.

If the curvature at the point of reaching the yield moment $\varkappa_y = M_y/(EI)_r$ is relatively small compared with the total curvature at failure, then rigid-plastic behavior can be assumed to simplify matters.

1.2.2 Limit State Theorems of the Theory of Plasticity

All methods for determining the collapse load using the theory of plasticity are based on the limit state theorems. They are derived here using, for illustration purposes, a simple statically indeterminate system and are established purely physically.

In Fig. 1.5 the elastic and plastic solutions are summarized for a beam fixed at one end with two concentrated loads at the one third points. In the following, a statical and a kinematical method are investigated for obtaining the plastic solution.

(a) Static Solution

For the solution only statical considerations are employed.

- Equilibrium conditions for the beam:

$$\frac{dM}{dx} = V \tag{1.1}$$

$$\frac{dV}{dx} = -q \tag{1.2}$$

- Statical boundary conditions for the beam (Fig. 1.5):

$$M_D = 0 \qquad \text{for boundary } x = \ell$$

- Yield Condition:
  Each solution for $M(x)$ is admissible which fulfils the equilibrium conditions and the statical boundary conditions, restricted however by the yield condition:

$$-M_R' \leq M(x) \leq M_R \tag{1.3}$$

That is, the moment is bounded by the negative value, $-M_R'$, and the positive value, M_R, of the plastic resistance moments of the beam.

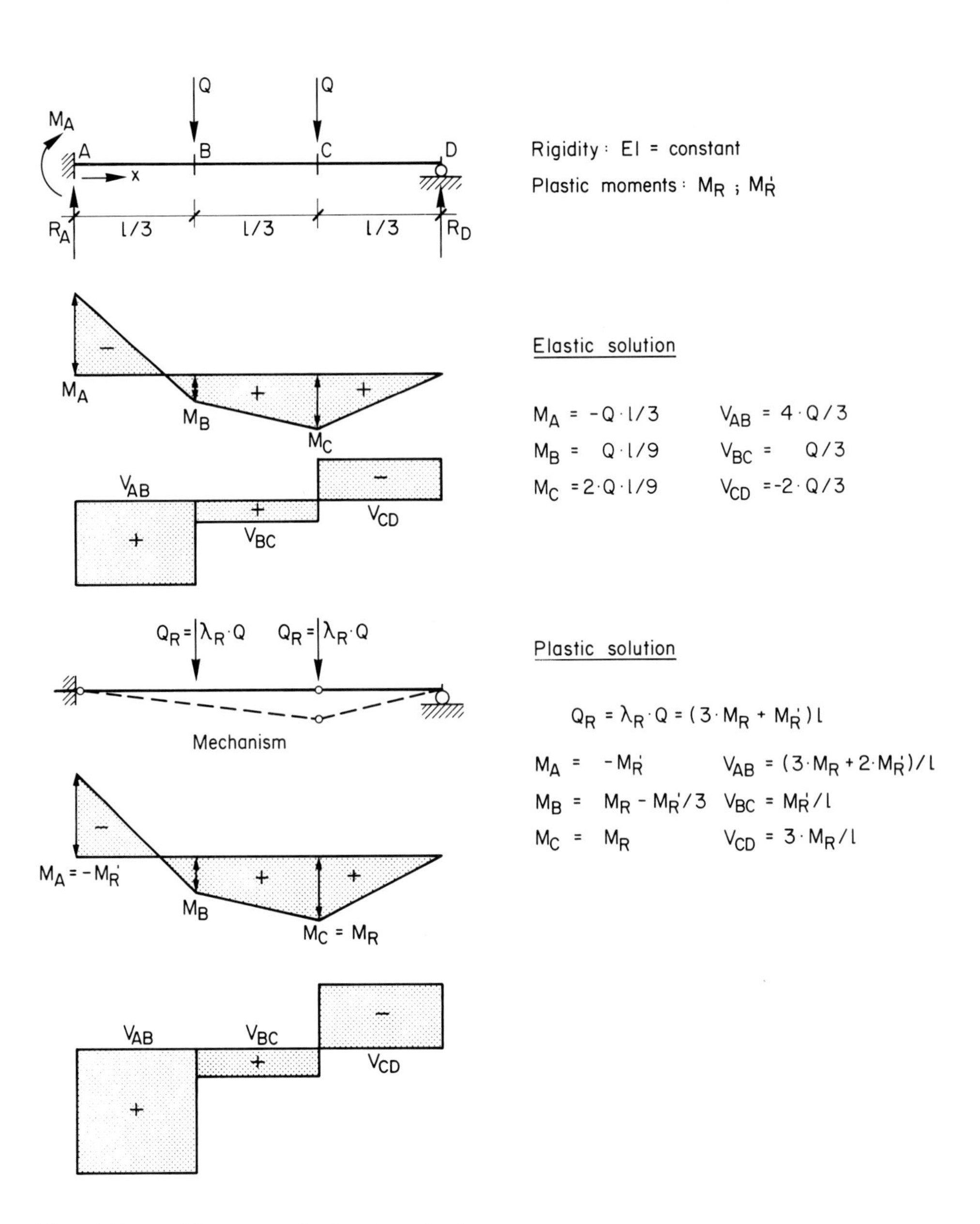

Fig. 1.5: Elastic and plastic solution

Three statically admissible equilibrium systems are shown in Fig. 1.6. From these it is evident that in general a system in equilibrium, which fulfils the statical boundary conditions and the yield conditions, still does not necessarily produce a mechanism.

In order for a mechanism to develop it is necessary to introduce some statical weakness, e.g. by removing supports or end fixity, by artificially weakening the cross-section by means of hinges ($M = 0$), etc. For this reason, loads which correspond to such equilibrium systems lie below the collapse load. Thus it follows that:

Lower Bound Theorem $[Q_S] \leq [Q_R]$:

"*A load system* $[Q_S]$, *based on a statically admissible stress field which nowhere violates the yield condition is a lower bound to the collapse load* $[Q_R]$."

Here the following assumptions and definitions have to be observed:

- A condition is proportional loading, i.e. all loads remain proportional to one another or $[Q_S] = \lambda_S \cdot [Q]$. Thus the whole loading system may be controlled by one parameter. If this is not the case, then the load combinations have to be investigated individually.

- A stress state is said to be statically admissible if it fulfils the equilibrium conditions and the statical boundary conditions.

- A mechanism is described by the state whereby plastic regions have developed to the extent that the system can further deform without increasing the load. The plastic zones take the form of hinges or hinge lines.

- A statically admissible stress state, which does not violate the yield condition, exhibits in general an insufficient number of hinges to be a mechanism. In order to actually form a mechanism a weakening of the system is necessary (see Fig. 1.6). Only the maximum value of $[Q_s]$ corresponds to a mechanism:

$$\text{Maximum } [Q_S] \Rightarrow [Q_R] \tag{1.4}$$

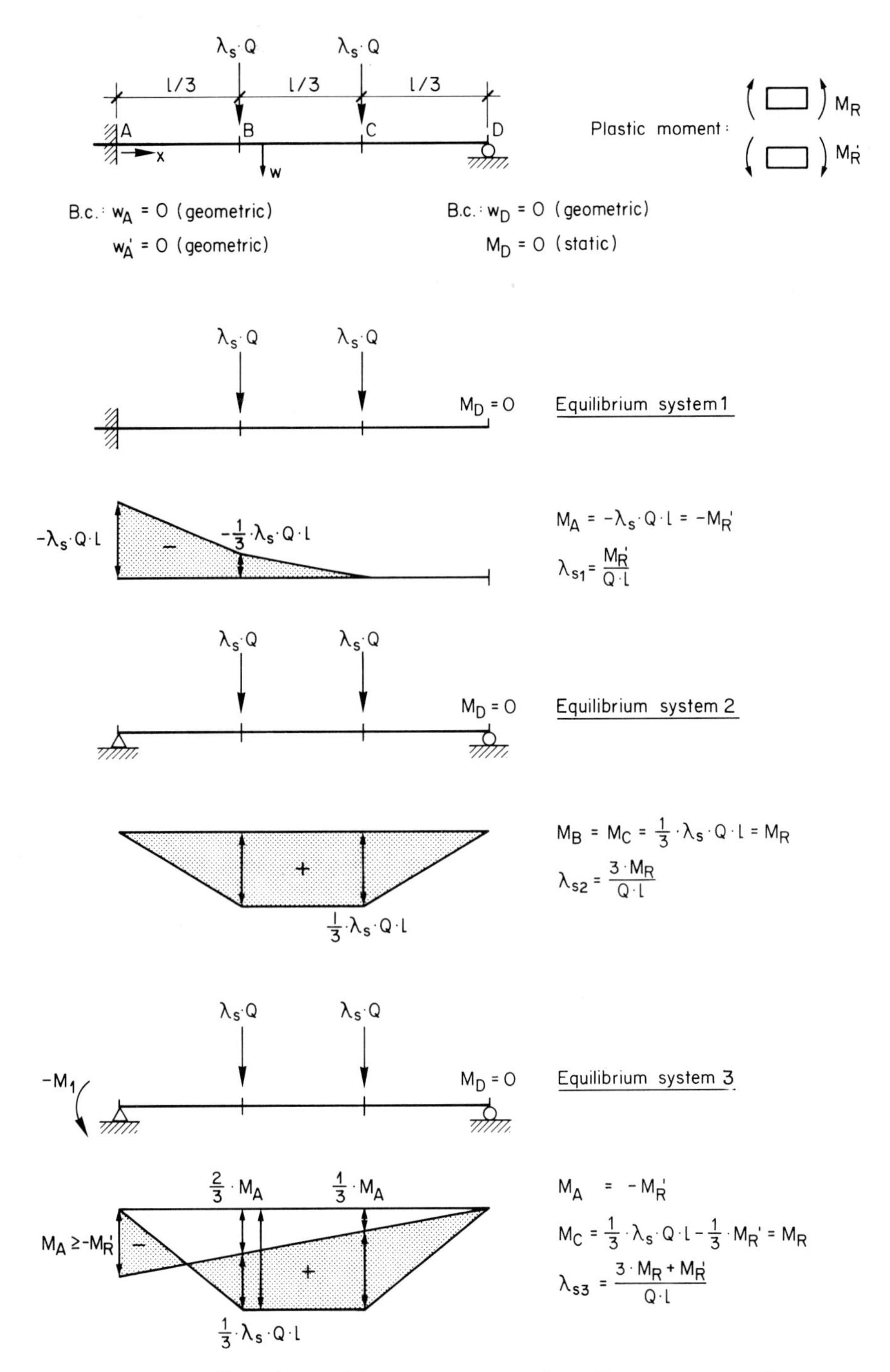

Fig. 1.6: Statically admissible stress states (bending moment diagram)

(b) Kinematic Solution

The dual approach is based on purely kinematical considerations. The assumed mechanism fulfils the geometrical boundary conditions. The hinge rotation is governed by the yield condition and the flow rule. All other structural components are considered to be rigid.

For illustration purposes mechanism 1 (Fig. 1.7) is explained. Assumed are a negative plastic moment $M_B = -M_R'$ at hinge B and a positive plastic moment $M_C = M_R$ at hinge C. The parts *B-C* and *C-D* form a mechanism. The system is in equilibrium with the loads Q_{K1}. Since the mechanism is determine by one degree of freedom, angle $3 \cdot \vartheta$, equilibrium can be expressed most easily by the principle of virtual work.

Principle of virtual displacement or virtual work:

If an equilibrium system is subjected to a virtual displacement, the sum of the external and internal work is zero:

$$A_a + A_i = 0 \tag{1.5}$$

The virtual displacements are fictitious displacements which are so small that they do not affect the equilibrium state. If the principle is expressed in terms of velocities, the magnitude of the virtual velocities has no influence on the equilibrium state.

In Fig. 1.7 equilibrium is formulated in each case by a global statement based on the principle of virtual displacements. Concerning the internal work A_i it should be mentioned that in rigid-plastic systems it is only expended in the plastic hinges. It is always of a dissipative form and is therefore negative:

$$A_a = -A_i \tag{1.6}$$

All contributions to the right handside are always positive.

The situation in Fig. 1.7 may be described as follows. Corresponding to mechanism 1 there is a bending moment diagram that severely violates the yield condition in part *A-B* of the beam, i.e. $M < -M_R'$. The system could only support the calculated load Q_{k1} if part *A-B* were correspondingly strengthened. The same applies for mechanism 2 which causes violation of the yield condition in the middle part of the beam.

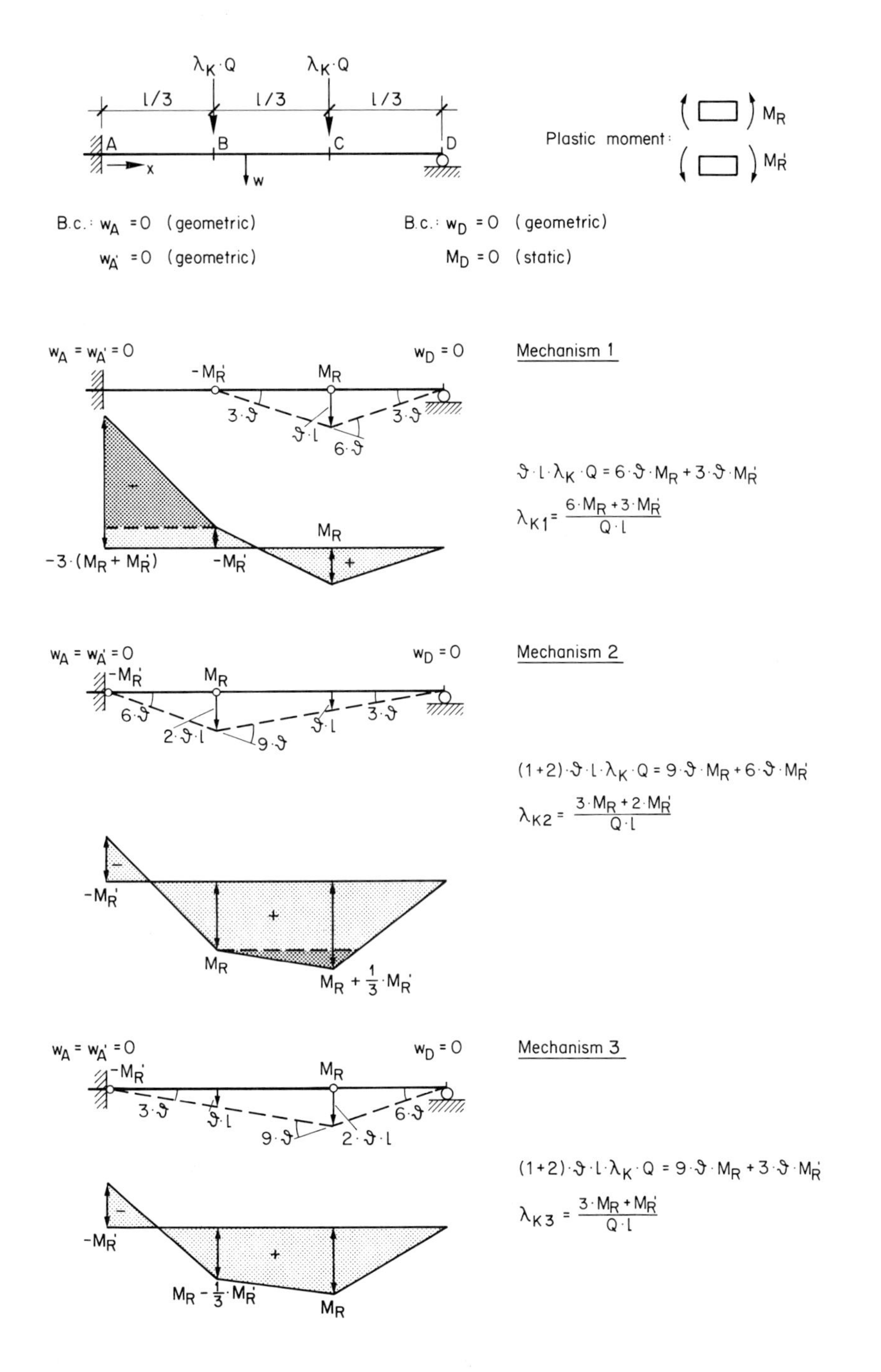

Fig. 1.7: Kinematically admissible displacement fields (mechanisms) with associated bending moment diagrams

Mechanism 3, however, exhibits a bending moment diagram which nowhere violates the yield condition, i.e. $-M_R' < M(x) < M_R$. At the same time there are two plastic hinges, $M_A = M_R'$ and $M_C = M_R$. Thus a further increase of load is not possible, and it follows that:

Upper Bound Theorem $[Q_K] \geq [Q_R]$:

"*A load system* $[Q_K]$*, which is in equilibrium with a kinematically admissible displacement field (velocity field) forming a mechanism is an upper bound to the collapse load* $[Q_R]$."

The following definitions apply:

- A displacement field (velocity field) is kinematically admissible if the resistances and deformations correspond to the yield condition and the flow rule and if the geometrical boundary conditions are fulfilled.

- A mechanism is a kinematically admissible displacement field and exhibits one degree of freedom.

A mechanism leads in general to a state of stress which violates the yield condition in parts of the structure. In order to be able to support the corresponding load $[Q_K]$ the system would have to be strengthened in these parts. Only the minimum of $[Q_K]$ actually corresponds to a mechanism, for which the yield condition is not violated in any part of the system.

$$\text{Minimum } [Q_K] \Rightarrow [Q_R] \tag{1.7}$$

Fig. 1.8 shows the three conditions, which have to be fulfilled for a complete solution. The lower bound theorem violates, in general, the conditions for a mechanism and the upper bound theorem violates the yield condition.

Attention should be drawn to the fact that static or kinematic solutions are not approximate solutions. The computation of the collapse load uses the method of linear mathematical programming, which is based on linear equilibrium conditions and the yield conditions in the form of linear restrictions (inequalities). The maximum value of $[Q_S]$ corresponds to the collapse load $[Q_R]$. Each linear program possesses a dual solution $[Q_K]$, whereby the minimum value of $[Q_K]$ corresponds to the collapse load $[Q_R]$. The two limit theorems of the theory of plasticity represent the physical formulation of the mathematical problem.

condition	static solution	complete solution	kinematic solution
equilibrium	ok	ok	ok
yield condition	ok	ok	?
mechanism	?	ok	ok
result	lower bound $[Q_S] \leq [Q_R]$	collapse load $[Q_R]$	upper bound $[Q_K] \geq [Q_R]$
method	static method	———	mechanism method

Fig. 1.8: Overview of plastic solutions

The limit theorems are valid also for two and three-dimensional problems, in which the simplification of rigid-plastic material behavior is assumed.

1.3 ENGINEERING METHODS

For the practical application of the theory of plasticity engineering methods have been developed which are based on the limit theorems.

In the previous section the limit theorems of the theory of plasticity were introduced with the aid of a simple example. The statical system, the geometry and the plastic resistances of the structural element were given. The collapse load was determined by applying the theory of plasticity. This procedure corresponds to the method of calculating the collapse load of a given structure i.e. plastic analysis (Fig. 1.9a). In this case the kinematic (mechanism) method allows an estimate of the collapse load of a structure of given dimensions and reinforcement to be made in an efficient way. Since the kinematical approach leads to upper bounds the result is on the unsafe side and must be assessed with the help of adequate experience. Furthermore, the stress state is generally only known in the hinge zones, so that a check on components and details outside of these zones is not possible. Thus in certain cases a statical solution is necessary to achieve a reasonable bounding of the collapse load and to check important details.

However, if, for a given load case with a known load intensity, the reinforcement and the geometry of cross-section are sought, the problem becomes one of design (Fig. 1.9b). In this case it is advantageous to apply the statical method. Firstly, the internal stresses are determined. The reinforcement is then proportioned in such a way that the yield stress is nowhere violated. It is convenient to reformulate the lower bound theorem as follows:

> "*In a plastic design a stress field is chosen such that the equilibrium conditions and the statical boundary conditions are fulfilled. The dimensions of cross-section and the reinforcement have to be proportioned such that the resistances are everywhere greater than or equal to the corresponding internal forces.*"

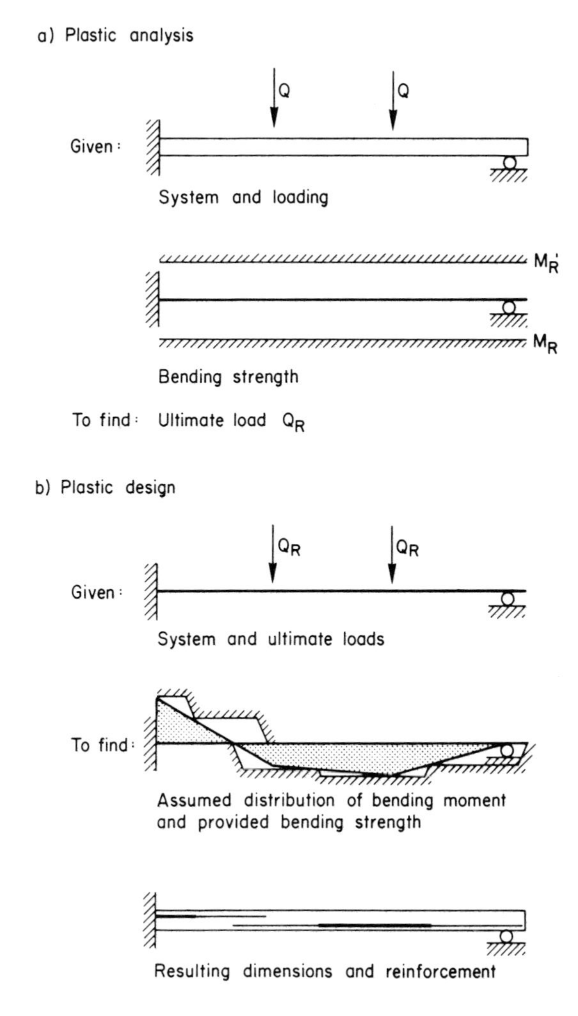

Fig. 1.9: Verification and design

In the design of framed structures instead of selecting a stress field one has to select a distribution of stress resultants, i.e. internal forces and moments. The admissible stress state corresponding to a system of loads determines the resultants in all parts of the structure. Thus all structural components and, especially important, details can be accurately proportioned and designed.

Fig. 1.9b illustrates the plastic analysis and design methods, in the case of collapse as the governing design criterion.

2 STRESS FIELDS FOR SIMPLE STRUCTURES

2.1 INTRODUCTION

In this chapter the behavior and strength of simple structures made of reinforced or prestressed concrete is investigated with the aid of stress fields. In particular, the webs and flanges of beams, simple walls, brackets, bracing beams and joints of frames are investigated. By this means, the majority of design cases are already covered.

In reality, all structural components are three-dimensional. Here, however, components are considered either directly as two-dimensional plate elements (i.e. the plane stress condition with no variation of stress over the thickness of the element) or they are subdivided into several plates. Since two-dimensional structural elements are statically redundant, it is possible for a particular loading to be in equilibrium with many (theoretically an infinite number of) stress states. If the lower bound method of the theory of plasticity is employed, then an admissible stress field or any combination of such stress fields may be selected. In chapter 4 it is shown that this method is suitable for the design of reinforced concrete structures, and the consequence of the choice of the final structural system on the structural behavior is dealt with in detail.

The first cases of the use of this method date back to Ritter [6] and Mörsch [4], who already at the beginning of the century investigated the resultants of the internal stresses by means of truss models. Stress fields based on the theory of plasticity were developed by Drucker [7] in the early sixties. Excellent summaries of the further development of this method are to be found in the publications of Thürlimann et al. [8], [9], Nielsen [10], Chen [11], Müller [12], Marti [13] and Collins, Mitchell [14]. A related application of the method, influenced by elasticity considerations, was proposed by Schlaich et al. [15], [16].

2.2 BEAMS WITH RECTANGULAR CROSS-SECTION SUBJECTED TO BENDING AND SHEAR

Although the behavior of these structural elements is readily understood or at least qualitatively known, emphasis is given here to the development of stress fields and not merely to the presentation of known solutions. In this way working with stress fields is introduced.

First of all a simply supported deep beam under two symmetrical point loads is considered. The structural behavior (i.e. distribution of internal forces) is described by means of a stress field. Subsequently, the external load is stepwise distributed and the slenderness of the beam increased, so that stress fields for simple beams of varying slenderness are obtained.

2.2.1 Deep Beams, Concentrated Loads

Given is a reinforced concrete beam of rectangular cross-section whose dimensions and loading at ultimate limit state are shown in Fig. 2.1. It should be noted that in accordance with modern structural codes the beam should be designed under factored loads Q_d using design resistances $f_{cd} = f_{ce}/\gamma_R$ and $f_{syd} = f_{sy}/\gamma_R$.

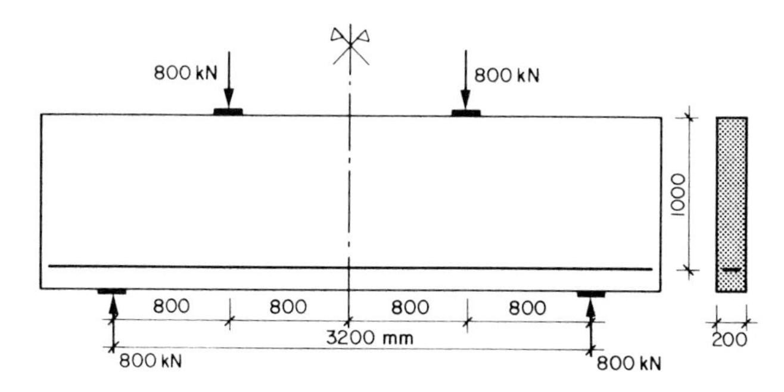

Fig. 2.1: Reinforced concrete beam with dimensions and loading

The strengths of the materials are:

- effective concrete strength f_{ce} = 20 MPa
- plastic limit of the reinforcing bars f_{sy} = 460 MPa

The effective concrete strength is dependent on the geometry, the type of stress field, confinement by reinforcement, etc., such that an appropriate effective strength has to be selected. For more details reference should be made to chapter 3.

The distribution of the internal forces can be obtained immediately by means of intuitive considerations. The distance of the load from the support is so small that it may be assumed to be supported directly. Thus, in effect, between the load and the support a strut under compression is formed. Fig. 2.2 shows the position of the resultants of the concrete struts, as well as the other compressive and tensile forces, which are necessary to maintain equilibrium.

The horizontal tensile force is carried by the reinforcement, while the horizontal compressive force is carried by the concrete.

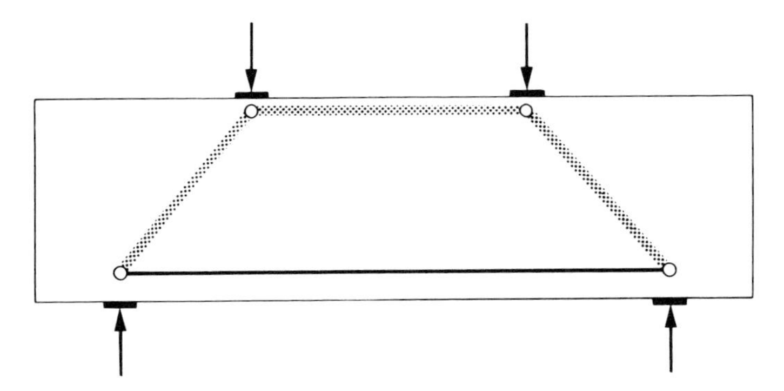

Fig. 2.2: Direct support, resultants of the internal forces

Statically equivalent to the horizontal compressive force there is a uniaxial stress field (compression strut) with constant stress intensity f_{ce}. This stress field is shown in Fig. 2.3a for an assumed lever arm.

Since between the compression strut and the upper surface of the beam there is a stress free zone the lever arm may be increased. Thus the optimum position of the strut can be determined iteratively. Fig. 2.3b shows the stress field after the first iteration, which may be taken as the solution. The middle part of the beam is in a state of pure bending. It follows that the iteratively obtained dimensions and forces could also have been determined analytically by conventional methods.

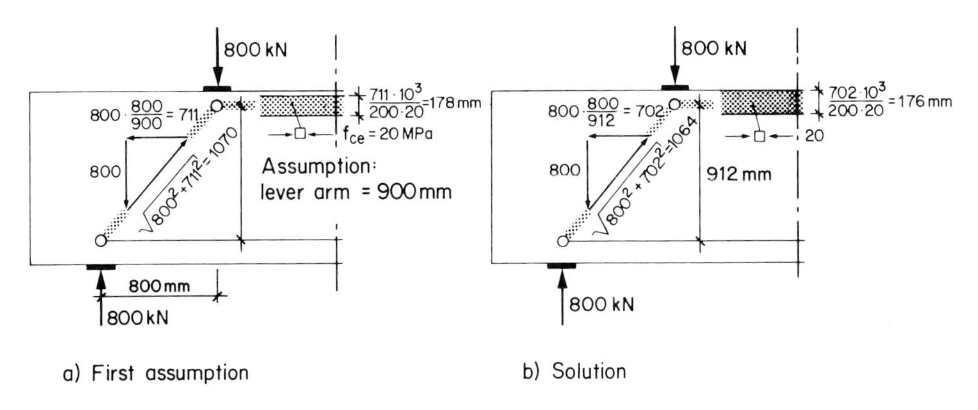

Fig. 2.3: Development of the compression strut

The solution presented above with constant stress intensity in the strut is only one of the possible solutions according to the lower bound theorem of the theory of plasticity. On the inner boundary of the strut there exists a discontinuity in the state of stress. At this discontinuity the stress intensity falls abruptly from f_{ce} to zero (Fig. 2.4a).

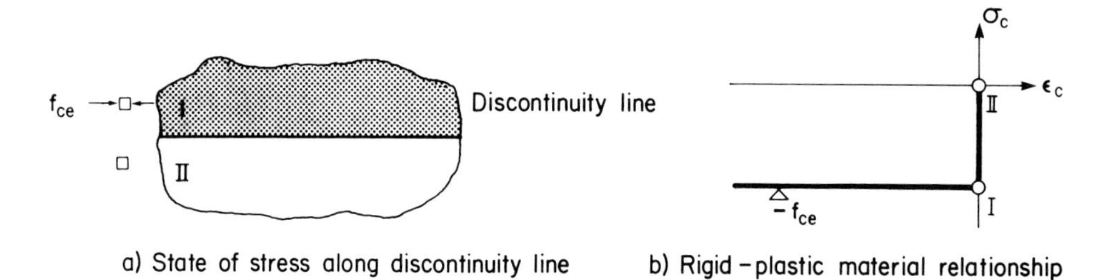

Fig. 2.4: Line of stress discontinuity

Such a discontinuity is only admissible for a rigid-plastic material relationship (Fig. 2.4b). Solutions by means other than stress discontinuities, however, are also possible (Fig. 2.5).

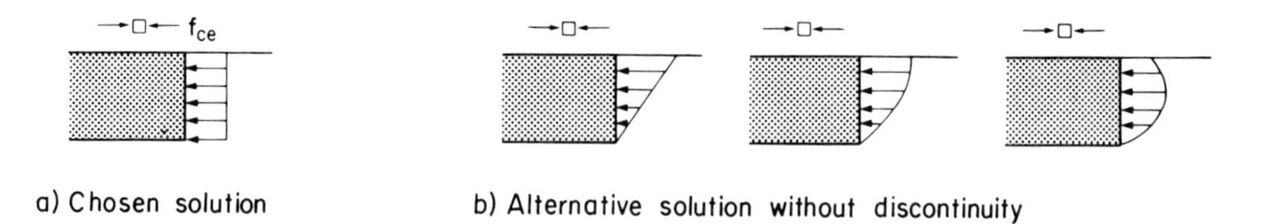

Fig. 2.5: Different stress distributions in the compression strut.

For typical amounts of reinforcing steel the choice of the stress distribution has no great influence on the position of the resultants. Therefore, it is best to select the simplest stress distribution possible. For beams with high amounts of longitudinal reinforcement or considerable normal forces a stress distribution according to Fig. 2.5b should be chosen. Such cases are treated in section 4.4.

A strut with constant intensity can also be used to describe the state of stress between the load and the support. The width of the strut results from the force in the strut divided by the effective concrete strength and the thickness of the beam (Fig. 2.6). The same procedure is used to determine the effective width where the load is introduced and at the support.

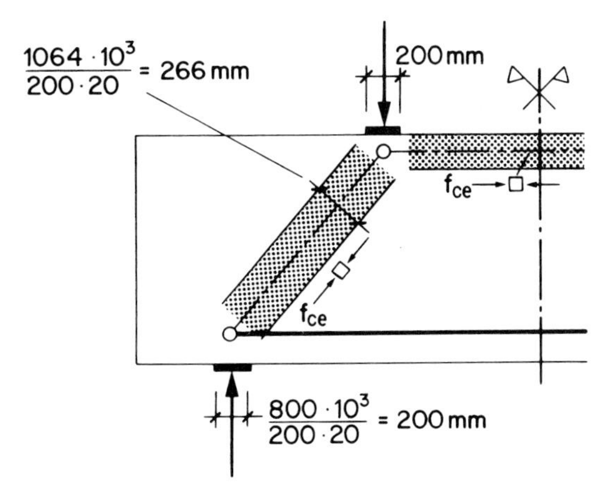

Fig. 2.6: Compression strut and compression diagonal

The development of the stress field in the joint region is shown in Fig. 2.7. Thus in the joints there is a biaxial state of stress with both principal stresses equal to f_{ce}.

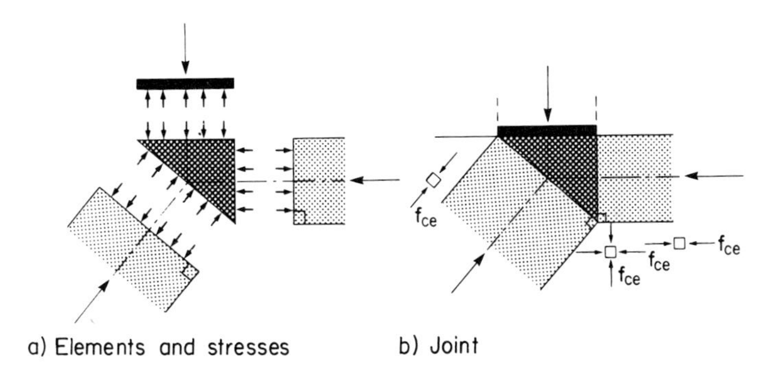

Fig. 2.7: Stress field of a joint

It should be noted that the discontinuity line between the strut and the nodal region is always normal to the strut if in the biaxially stressed region a state of isotropic stress ($\sigma_2 = \sigma_3$) exists. The joint region above the support is considered in an analogous way (Fig. 2.8a).

The state of stress in the biaxially stressed region is in equilibrium with the compressive force in the diagonal strut, with the support reaction and the force in the reinforcement. Both the force in the reinforcement and the support reaction are uniformly distributed over the associated discontinuity lines.

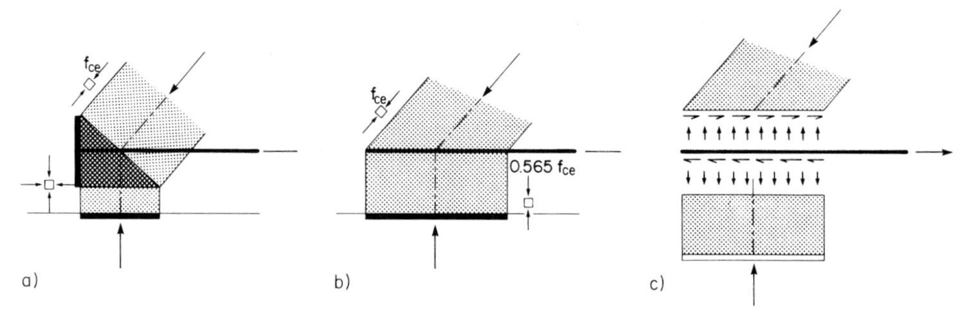

Fig. 2.8: Joint at support

According to this stress field the force in the reinforcement must act right up to the left side of the support. An alternative solution is shown in Fig. 2.8b; the force in the reinforcement in the region of the support is gradually reduced. The stress field is only admissible if the force in the reinforcement can be transmitted to the concrete within the region of the joint (Fig. 2.8c). It is clear that the area of the support is increased and the stress in the vertical compression field is correspondingly reduced. The stress fields for the whole beam are shown for both variants in Fig. 2.9.

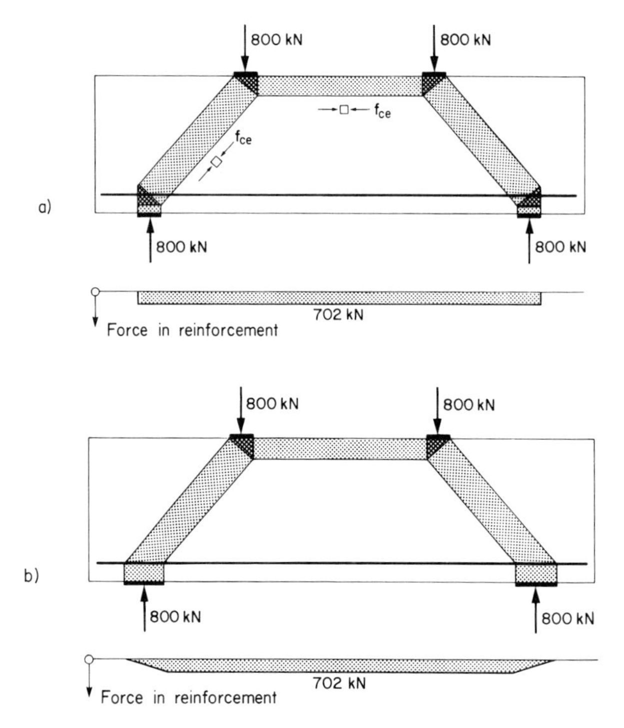

Fig. 2.9: Stress fields and force in the reinforcement

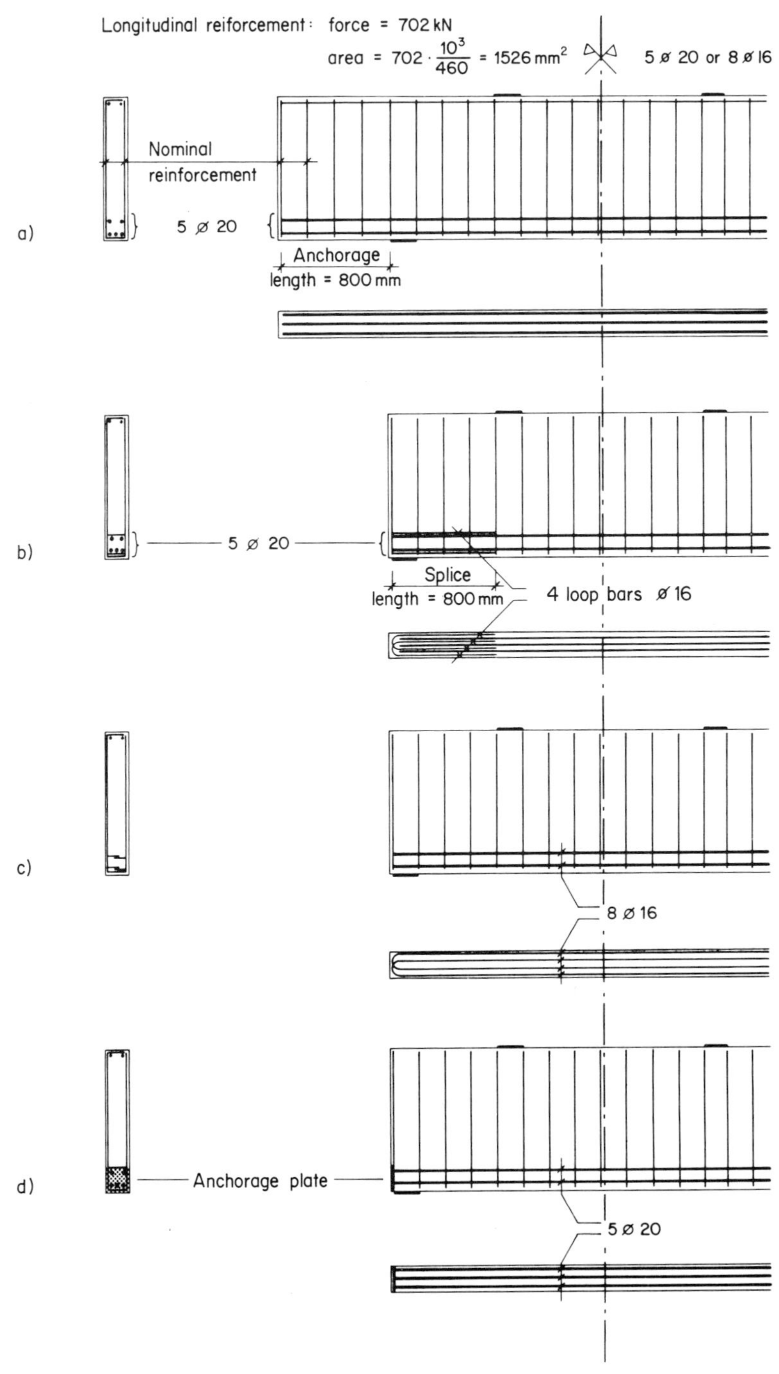

Fig. 2.10: Reinforcement sketches

A possible arrangement of reinforcing steel is shown in Fig. 2.10a, where the complete reinforcement is anchored behind the support. Figs. 2.10b to d show solutions whereby the anchoring of the reinforcement directly behind the support is achieved with the aid of stirrups, loops or anchor plates.

The results may be summarized as follows:

- Although the beam is subjected to high shear forces, from the statical point of view no stirrups are necessary.

- The distribution of the force in the reinforcement does not correspond to that of the moments. On the contrary, it is constant over the length of the beam, so that in the area of the supports the full force must be anchored.

Concerning the design of reinforced concrete structures in practice, even this example allows the following observation to be made: The stress field does not have to be developed quantitatively in all elements. By means of a qualitative sketch of the stress field an adequate distribution of the internal forces for an efficient design can be obtained, i.e. for determining the position and cross-sectional area of reinforcement, for the detailing as well as checking the stresses in critical regions.

2.2.2 Deep Beams, Several Concentrated Loads

The reinforced concrete beam of the previous example is considered further. The two loads of 800 kN are divided up into eight equal loads having the same resultant.

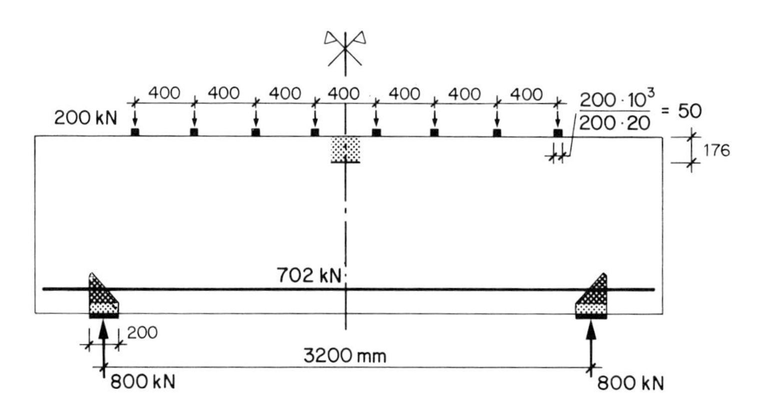

Fig. 2.11: Beam and loading

Since the latter is unchanged, the dimension of the horizontal strut in the middle of the beam and the reinforcement forces remain the same as in the previous example (Fig. 2.11).

The struts can be progressively developed. The development of the first strut is shown in Fig. 2.12a, the complete stress field is shown in Fig. 2.12b and the corresponding resultants with a Cremona force diagram are shown in Fig. 2.12c.

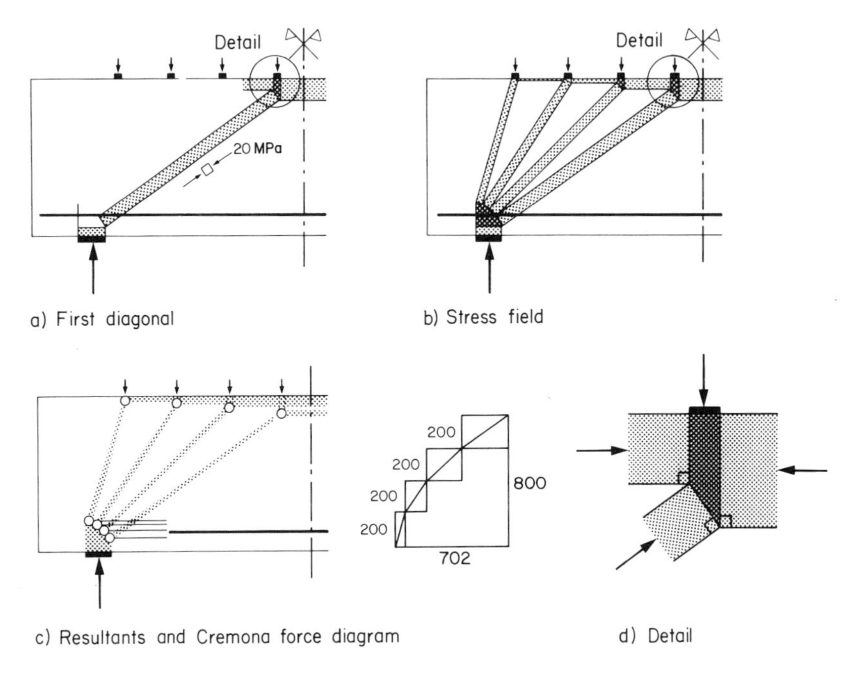

Fig. 2.12: Development of the stress field

2.2.3 Deep Beams, Distributed Load

The dimensions of the beam and the material properties are also retained for this example. Over the whole span a uniformly distributed load of 500 kN/m acts which corresponds to the failure load.

Since the loads produce the same resultants as in the first example, the dimension of the horizontal strut in the middle of the beam and the reinforcement force are unchanged.

The stress field can be developed in an analogous way to the previous example, in that the single load is replaced by an infinitely large number of concentrated loads (Fig. 2.13).

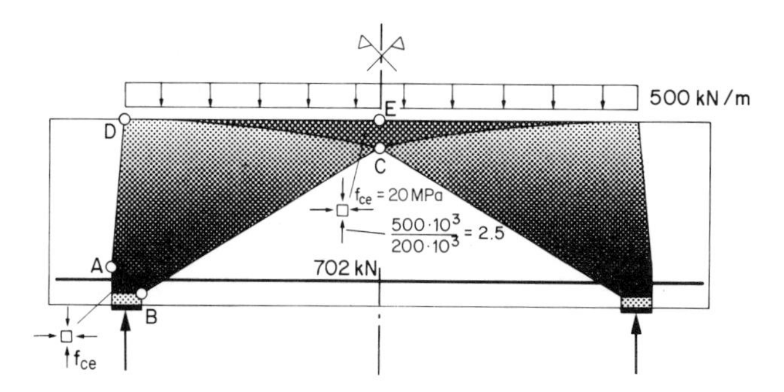

Fig. 2.13: Stress field

The stress field *ABCD* is called a fan. The stress intensity amounts to f_{ce} at the line of discontinuity *AB* and decreases hyperbolically in a radial direction. In the region *CDE* there is a biaxial state of stress. It can be shown that the discontinuity lines *AB* and *CD* are parabolic. The shape of these lines, however, is of no significance for practical design.

2.2.4 Beams with Medium Slenderness Ratio, Concentrated Loads

A beam with the same cross-section as in sections 2.2.1 to 2.2.3 is considered, but with the span doubled. The loads are still 800 kN (ultimate limit state) and act at one quarter of the span from the supports. The stress field developed in section 2.2.1 can be applied qualitatively for this case too (Fig. 2.14).

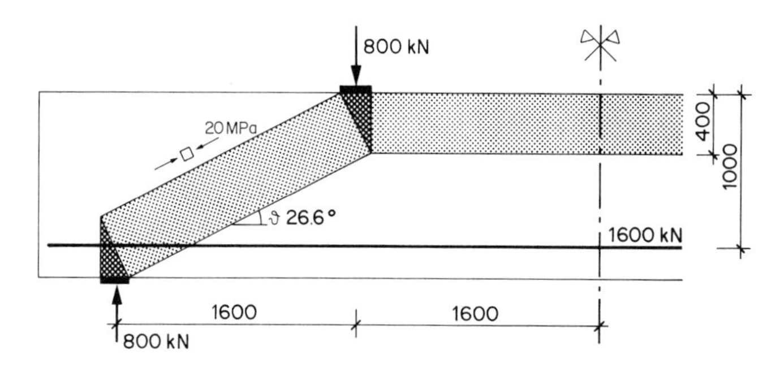

Fig. 2.14: Stress field for direct support

The thickness of the horizontal strut is 400 mm and the slope of the diagonal strut is 26.6°.

Experimental investigations have shown that this stress field can only develop under certain conditions. It will be shown in chapters 3 and 4 that force transmission in concrete can be problematic if an unreinforced compression strut is close to tension-stressed reinforcement over an extended length, i.e. if they intersect at a relatively small angle.

Thus, alternative solutions with more steeply inclined struts are sought. Fig. 2.15 shows how the problem can be solved by combining two structural systems.

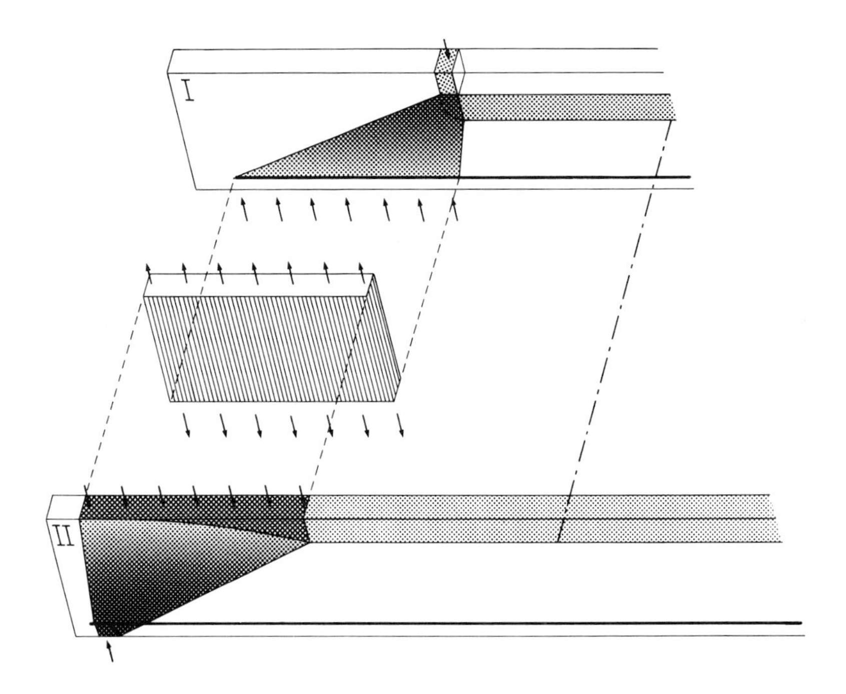

Fig. 2.15: Combination of two structural systems

The reactions of system I load system II by means of a vertical stirrup reinforcement, which must be well anchored both at the top and the bottom. The stress field of system II can be taken from that of section 2.2.3 (same loading and dimensions).

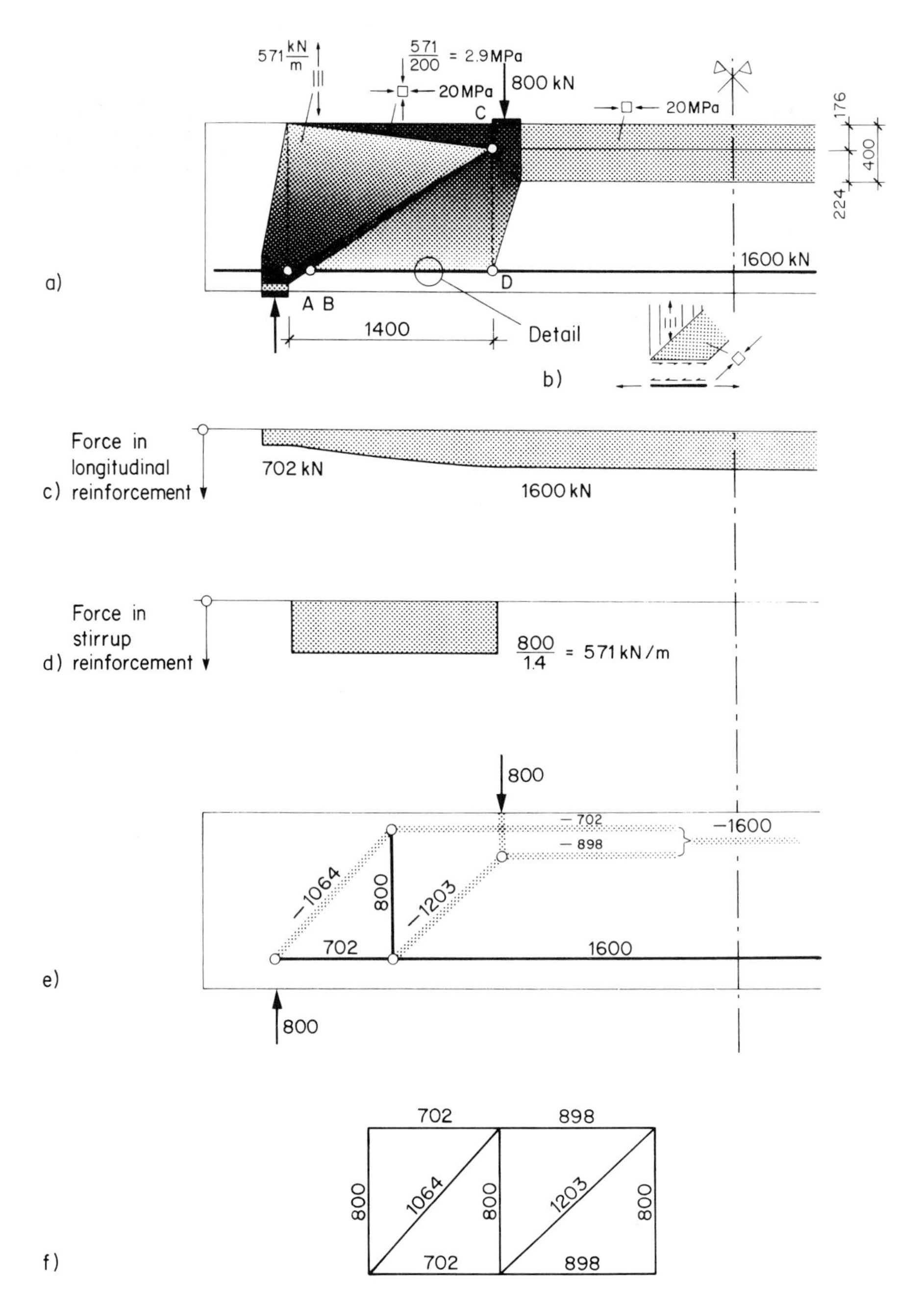

Fig. 2.16: Stress field (indirect support), forces in the longitudinal and stirrup reinforcement, resultants of the stress field, Cremona diagram

In developing system I it must be taken into account that the support reaction is distributed. A stress fan results, whose vertical component is resisted at the lower edge by the stirrups. The horizontal component is transmitted via bond stresses to the longitudinal reinforcement. Assuming a uniform distribution of force in the stirrups and due to a varying direction of principal stress in the fan it follows that there is a nonlinear variation of force in the longitudinal reinforcement. It should be noted that the whole height of the beam is not available in system I, since the upper zone of the beam is used for system II. In the middle of the beam the total thickness of the horizontal strut as well as the total force in the reinforcement are the same as in Fig. 2.14.

The complete stress field including all dimensions and forces is shown in Fig. 2.16.

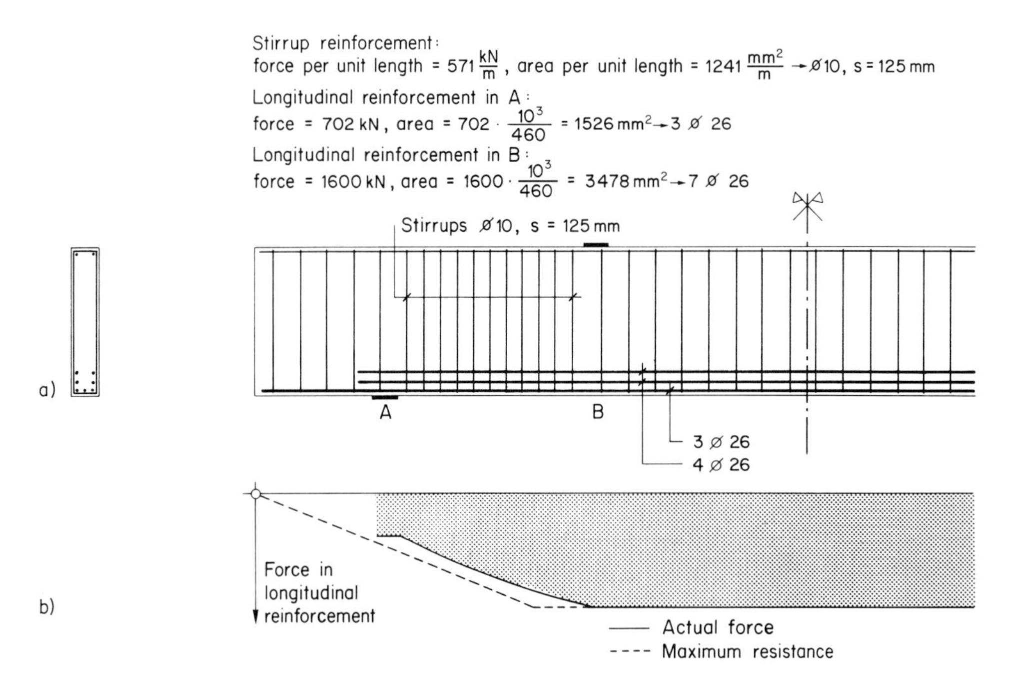

Fig. 2.17: Reinforcement sketch and force in longitudinal reinforcement

Fig. 2.17a shows a possible sketch of the reinforcement for this beam. Three longitudinal bars carry the support reaction and have to be fixed behind the support. The four lateral bars are loaded continuously by bond stresses. Fig. 2.17b shows a comparison between the actual force in the longitudinal reinforcement and the maximum resistance. The increase of the resistance can readily be obtained by assuming that the gradient corresponds to the bond strength of the reinforcement. It is evident from this figure that the four longitudinal bars do not have to be anchored behind the supports.

From Fig. 2.16a it is evident that in the region *ABC* the two fans interfere, which leads to a minor violation of the yield condition ($|\sigma| > f_{ce}$). Although this solution is adequate for practical purposes, an additional stress field is developed without a zone of interference (*ABC*). The stirrups are distributed over the length *BD*, which leads to a shift in their resultant. Thereby the position of point *C* is changed and thus also that of point *B*. It follows that an iterative process is required. The final stress field is shown in Fig. 2.18.

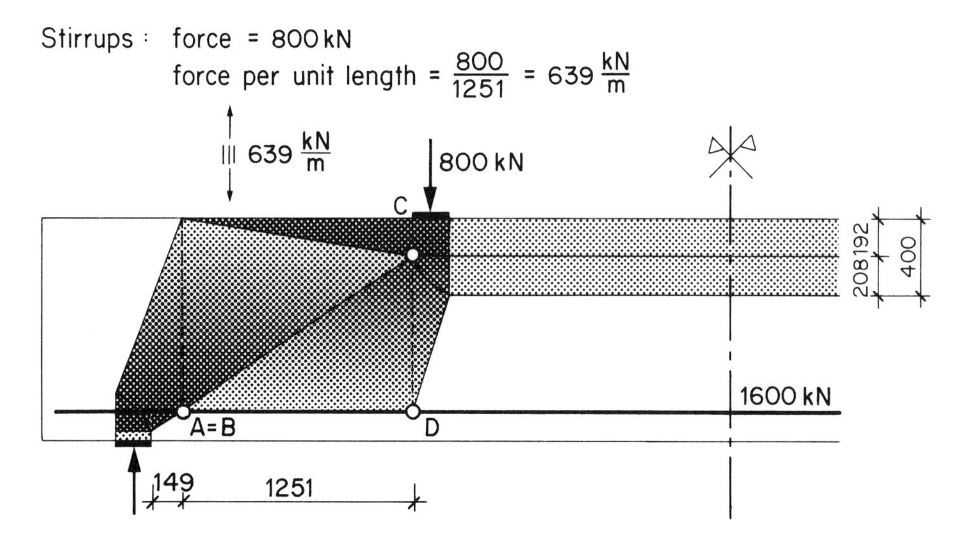

Fig. 2.18: Stress field

The superposition of the two stress fans can also be overcome by introducing the horizontal fan components at the support by means of the bond stress in the reinforcement according to Fig. 2.8b.

2.3 BEAMS WITH I-CROSS-SECTION SUBJECTED TO BENDING AND SHEAR

2.3.1 Slender Beams, Concentrated Loads

The span of the previous beam (Fig. 2.14) is doubled, the height, the material strengths and the loads remain unaltered. Apart from the change in the internal lever arm the force in the horizontal strut increases in proportion to the slenderness. Since in the previous example already 40% of the statical height was used, the cross-section must be enlarged by increasing the width in this region. As with the compressive force, the longitudinal reinforcement force also increases. In order to prevent the resulting large reinforcement area from leading to detailing problems, the cross-section in this area must also be enlarged (Fig. 2.19a).

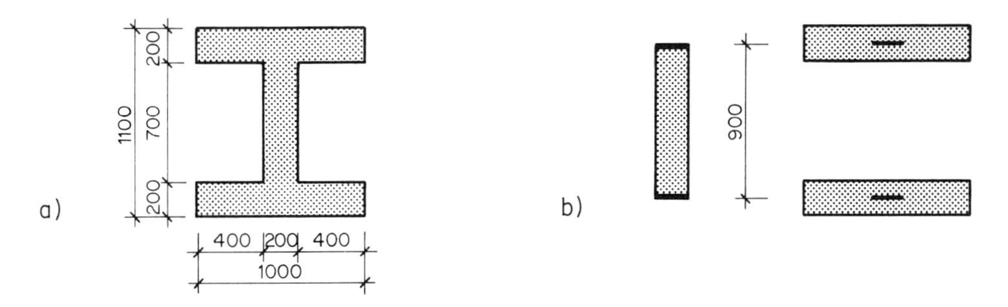

Fig. 2.19: Cross-section and components web, compression and tension flanges

It is assumed that the horizontal compression and tension resultants act at the center of gravity of the two flange plates. Thus the beam can be subdivided into three parts (Fig. 2.19b). The two flanges will be considered in sections 2.3.5 and 2.3.6.

The stress field describing the stressing of the web can be developed in an analogous manner to Fig. 2.15. The increased slenderness is taken into account in that three instead of two structural systems are combined such that the inclination of the diagonal compression field is not too small (Fig. 2.20.)

It should be observed that the statical height of the three structural systems remains constant. The horizontal compressive force spreads out in the compression flange.

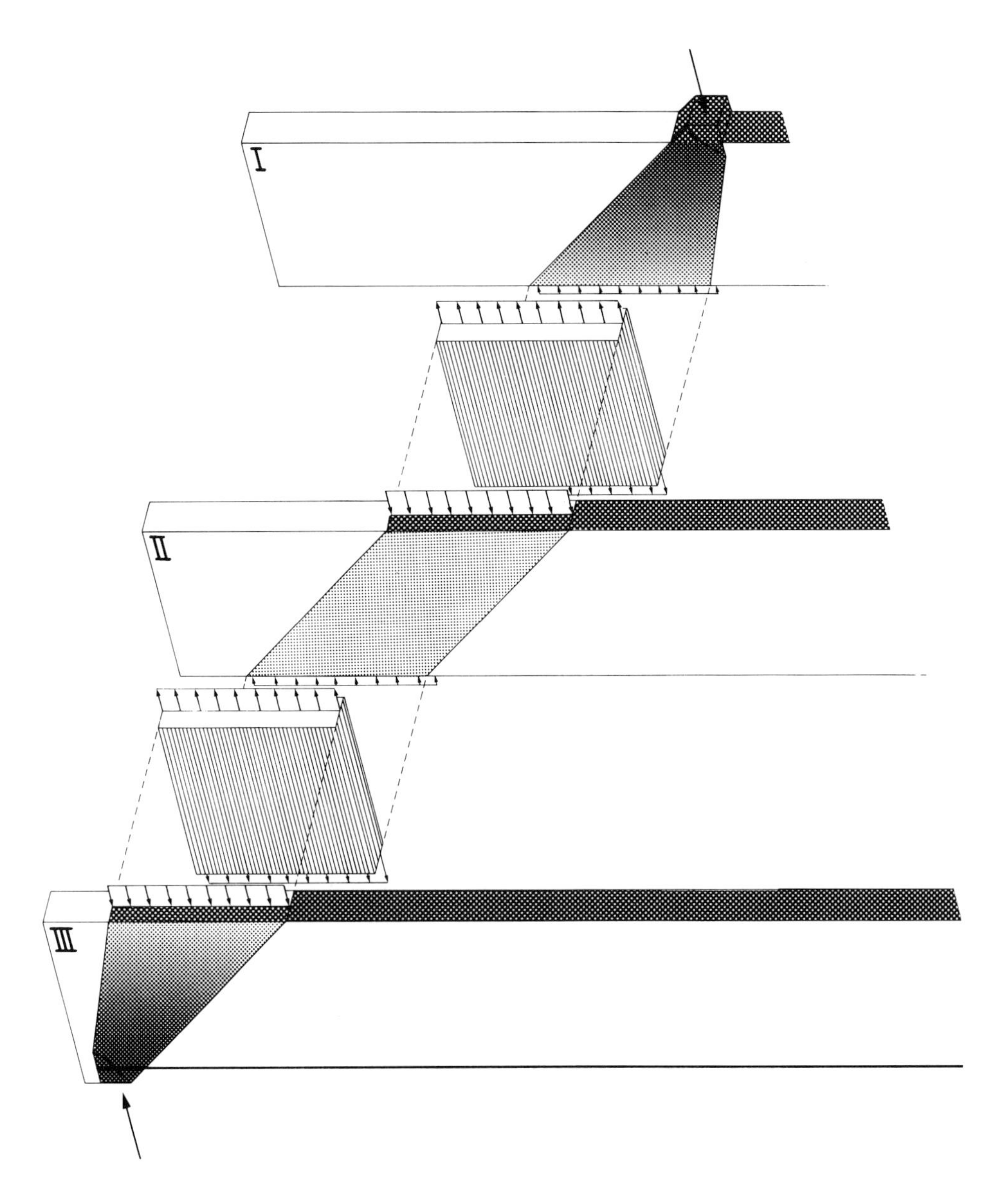

Fig. 2.20: Combination of three structural systems

Fig. 2.21 shows the stress field that results from the superposition of the three structural systems. The stress field in the concrete consists of two fans and an intermediate diagonal compression field. The stress in this strut is obtained by dividing the resultants by the width and thickness. Analogous to Fig. 2.16 there is a small interference of different regions, which for practical design purposes, however, is of no consequence. The problem could be solved as indicated in Fig. 2.18. The reinforcement is designed after the stress fields of both flanges are known.

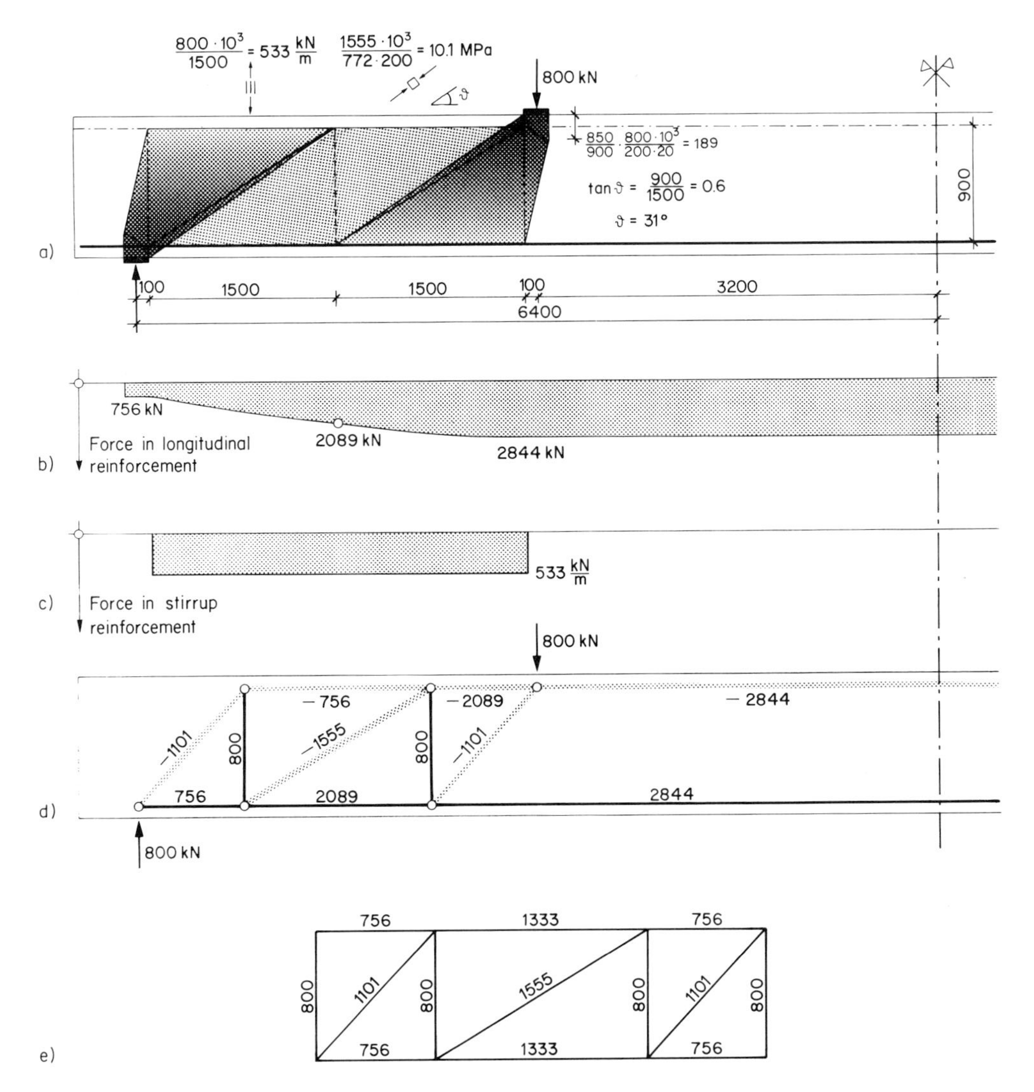

Fig. 2.21: Stress field, forces in the reinforcement, resultants of the stress field and Cremona diagram

As an alternative solution a stress field with triple suspension (combination of four structural systems) can be obtained, which causes a change in inclination of the strut. This stress field is shown in Fig. 2.22.

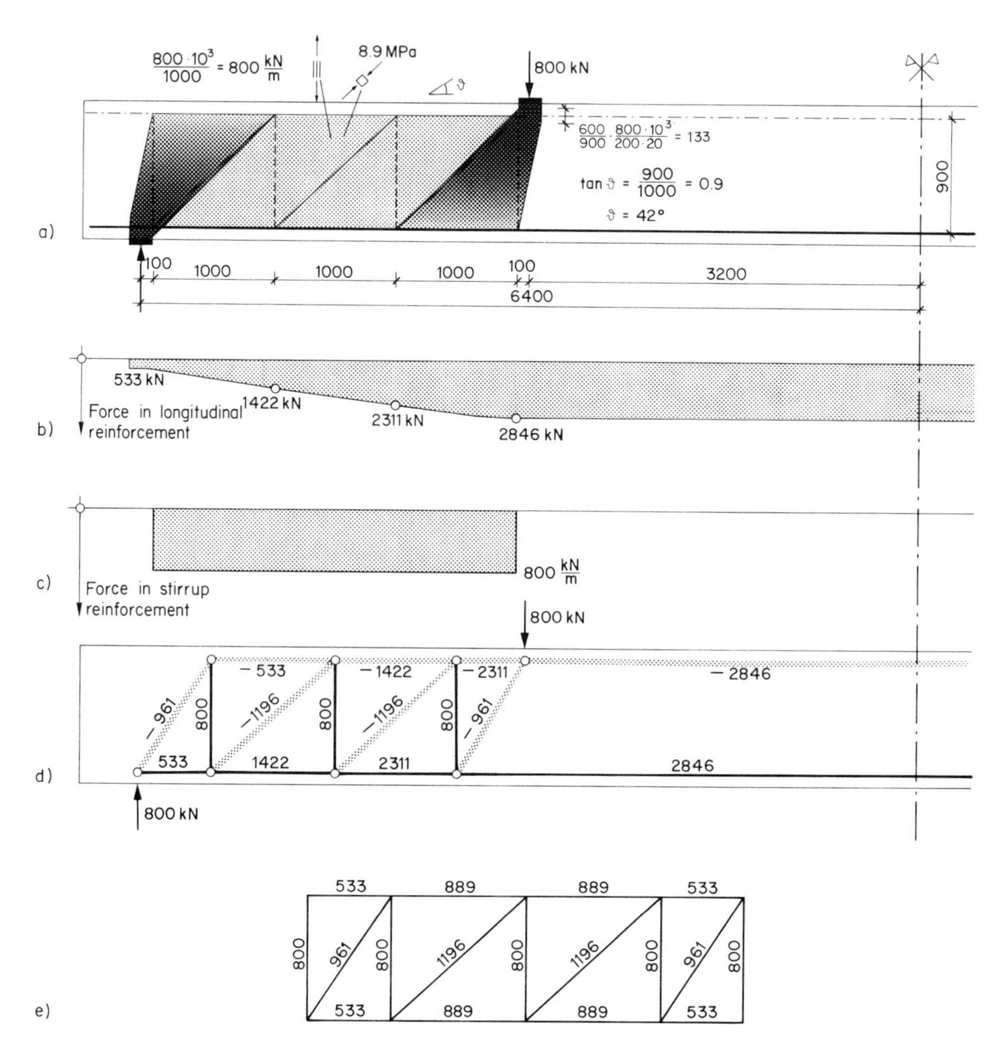

Fig. 2.22: Stress field, forces in the reinforcement, resultants of the stress field and Cremona diagram

The resultants of the stress field show that there is a basic underlying relationship between the solutions for examples with small, medium and large slenderness ratios. This relationship is schematically shown in Fig. 2.23.

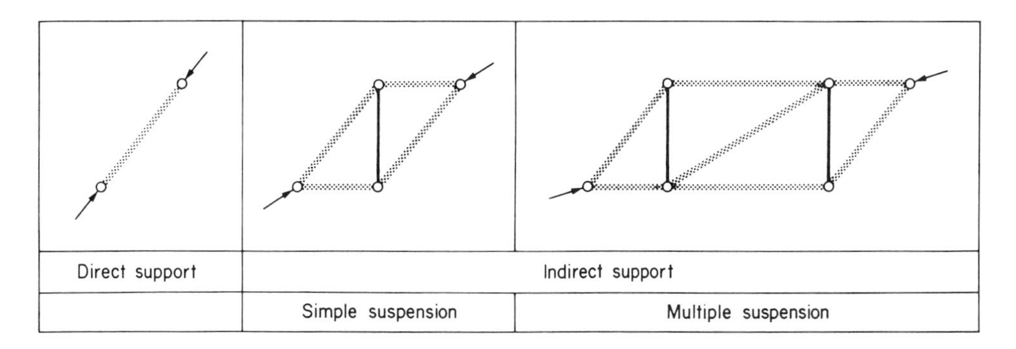

Fig. 2.23: Sketch of the structural behavior of the different examples

It should be observed that vertical equilibrium in the center part of the stress field (Fig. 2.22a) is achieved by superimposing two fields (Fig. 2.24). The vertical component of the inclined compression field is kept in equilibrium by the force in the stirrups. The components at the top and bottom boundaries are transmitted to the flanges.

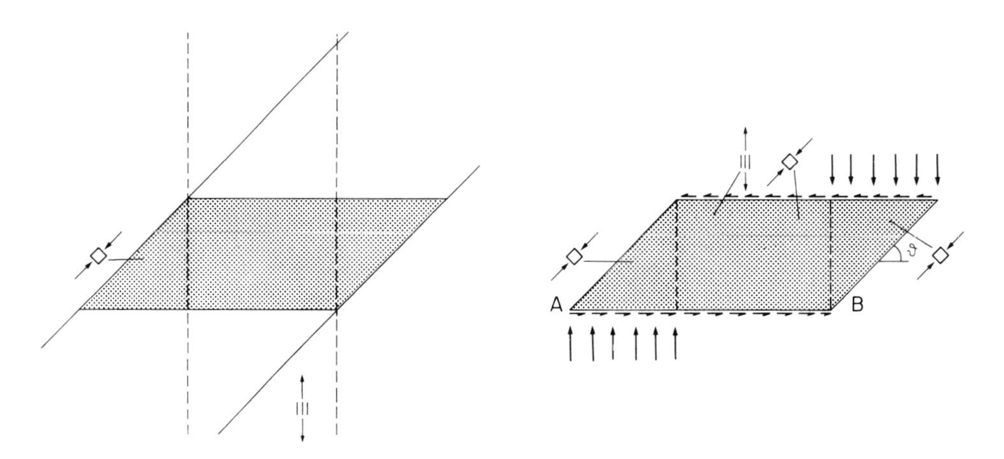

Fig. 2.24: State of stress in the web due to the superposition of two fields

The element shown in Fig. 2.24 can be developed independently of the slope angle ϑ and of the length *A-B*. Fig. 2.25 shows a stress field with tan ϑ = 0.75.

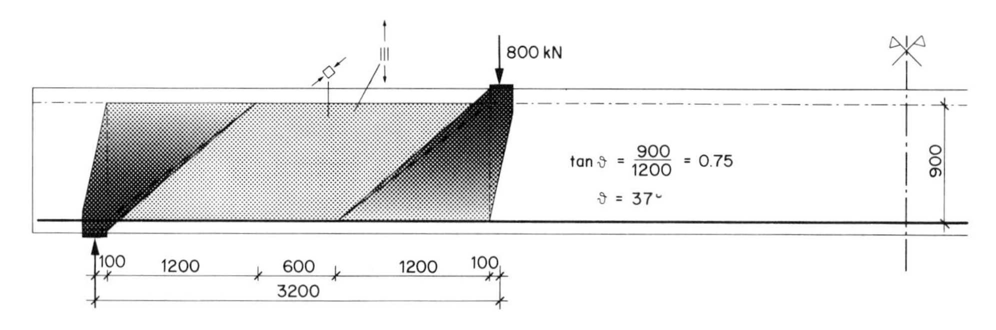

Fig. 2.25: Stress field

In the case of this stress field the associated resultants can no longer be represented in a simple way. However, for practical design purposes this is not necessary, since the internal forces can be determined at the lines of discontinuity using equilibrium considerations (Fig. 2.26).

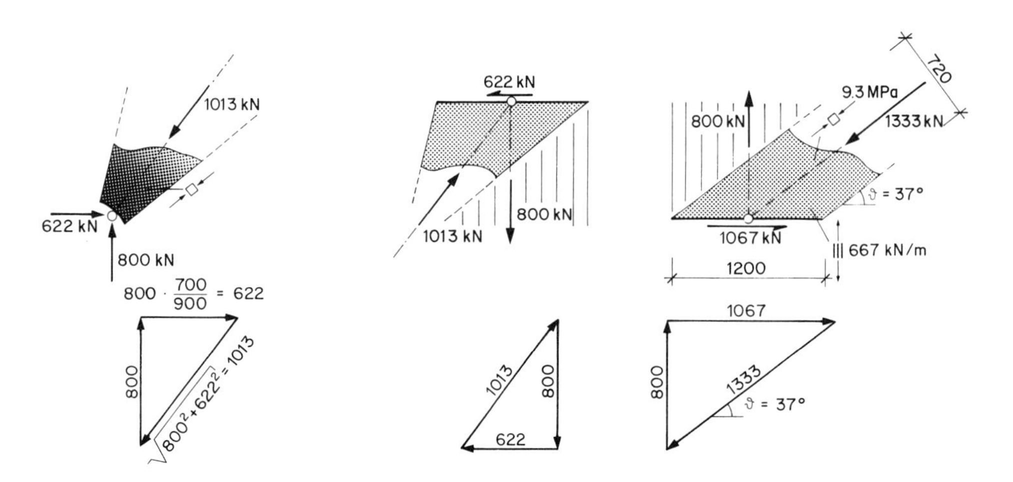

Fig. 2.26: Equilibrium at the discontinuity lines

The choice of the inclination ϑ of the compression field has an influence on the internal forces, as the three examples treated show. The most important values are shown in Fig. 2.27.

With increasing inclination of the compression field the horizontal reinforcement force decreases and the force in the stirrups increases. It is shown in chapter 4 that the choice of the distribution of the internal forces in a statically indeterminate system has an influence on the behavior under working load conditions. With the choice of the compression field inclination the behavior is influenced in an analogous manner.

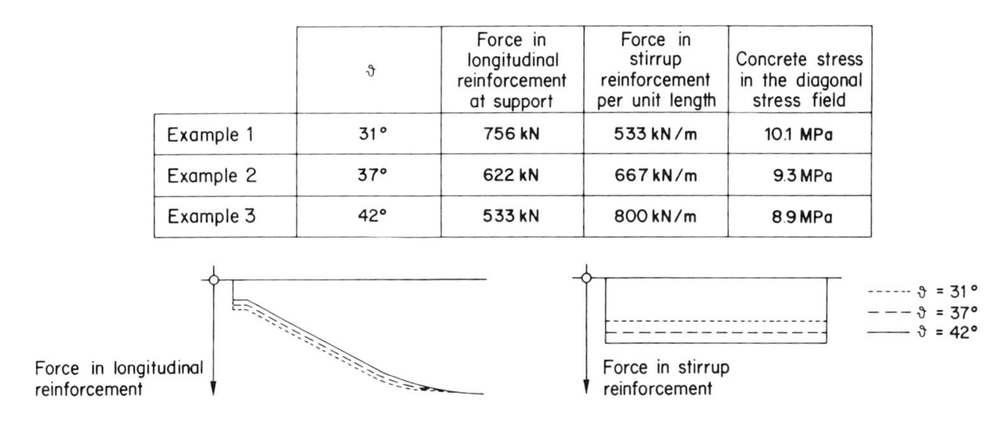

	ϑ	Force in longitudinal reinforcement at support	Force in stirrup reinforcement per unit length	Concrete stress in the diagonal stress field
Example 1	31°	756 kN	533 kN/m	10.1 MPa
Example 2	37°	622 kN	667 kN/m	9.3 MPa
Example 3	42°	533 kN	800 kN/m	8.9 MPa

Fig. 2.27: Internal forces for different compression field inclinations

The inclination of the compression field for optimum behavior with respect to serviceability (smallest possible redistribution of the internal forces from the working load to the ultimate load state) depends on several parameters (residual stress state, possible prestress or normal force, amount of reinforcement, stiffness of the compression flange, etc.) and is difficult to derive. Large departures from this solution have no marked influence on the serviceability. Solutions with either very large or very small compression field inclinations should, however, be avoided. A detailed discussion of this subject is given in section 4.1. In structural codes normally a range is given, in which a free choice is possible.

It is evident from Fig. 2.27 that the inclination of the compression field also has an influence on the intensity of the stress in the concrete. For highly stressed webs ϑ must be chosen so that the effective strength of the concrete is not exceeded. It should be noted that in this case the strength f_{ce} is considerably less than the cylinder strength. The state of strain in the web (strain resulting from the elongation of the reinforcement) and further interactions with the reinforcement are responsible for this reduction. These influences are discussed further in chapter 3.

2.3.2 Slender Beams, Distributed Load

A beam is considered which has the same dimensions as the one described in section 2.3.1. The two concentrated loads each of 800 kN are distributed over the whole length of the beam, so that a distributed load of 125 kN/m results.

The stress field for describing the internal stress state can be taken in part from Fig. 2.22a. As Fig. 2.28 shows, in this case there results a stepwise variation of the intensity of the inclined compression field and the forces in the stirrups over the length of the beam.

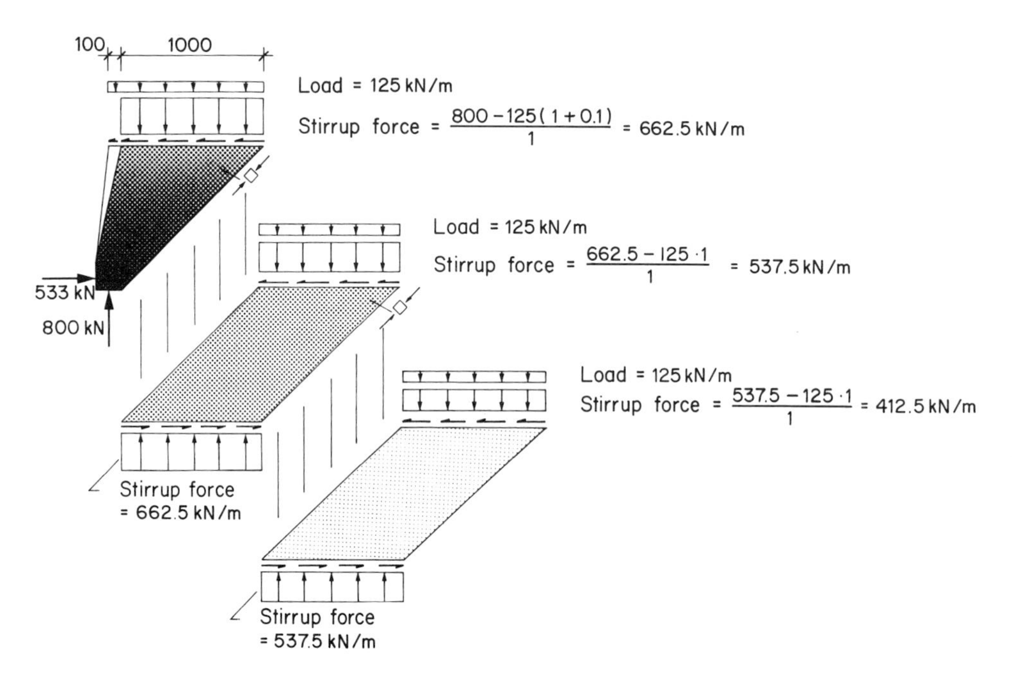

Fig. 2.28: Development of the stress field

In Fig. 2.29 the complete stress field, the reinforcement forces, the resultants of the stress fields and the Cremona diagram are shown. The force acting at the upper edge of the fan at the support plate (12.5 kN) is introduced directly. Thus at the left of this zone an additional small fan results, which is of no significance for practical design.

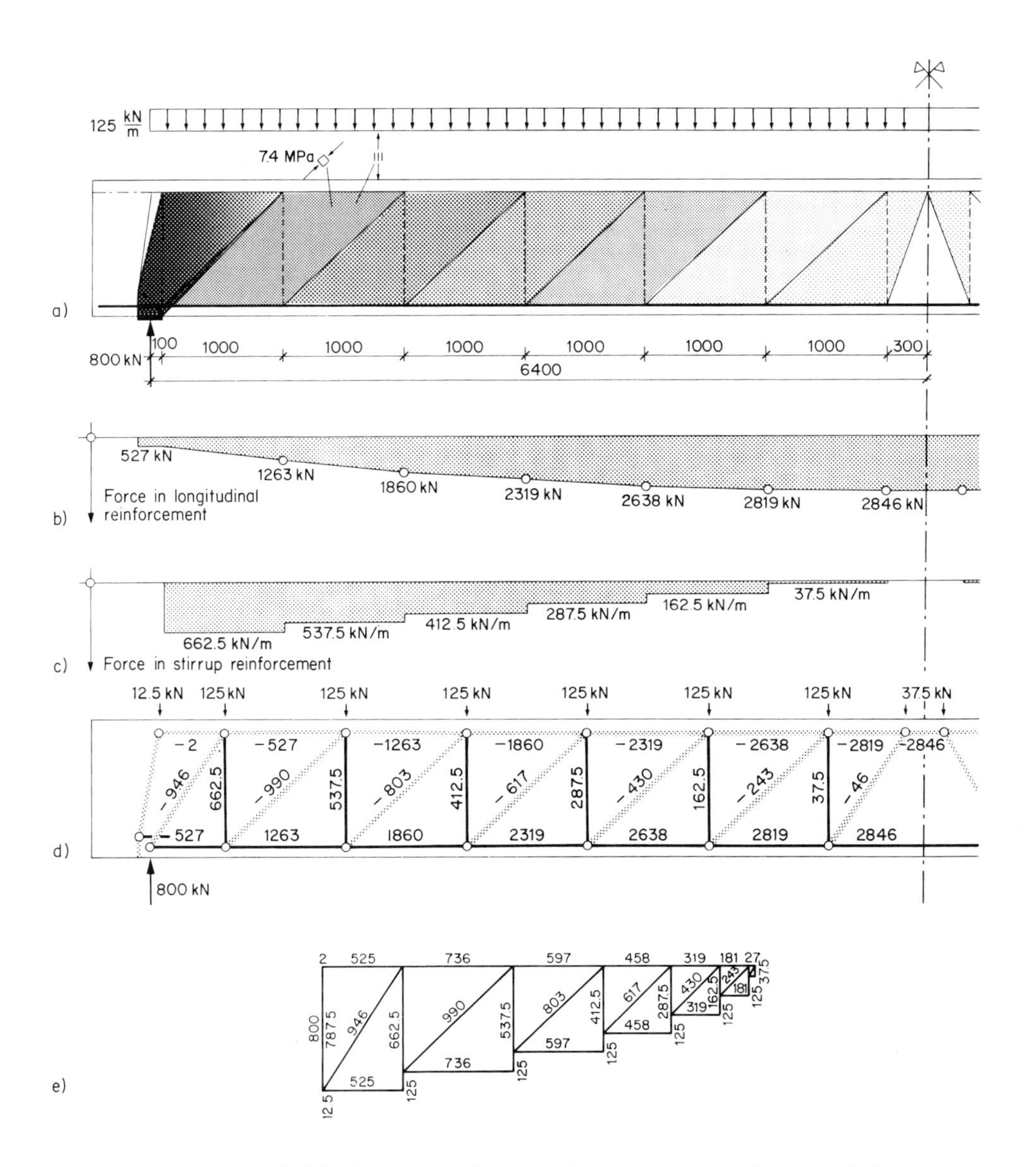

Fig. 2.29: Stress field, forces in the reinforcement, resultants of the stress field and Cremona diagram

In each case the inclination of the compression field can be varied over the length of the beam. Fig. 2.30 shows qualitatively a general solution.

The stressing of the flange plates as well as a sketch of the reinforcement for the complete beam will be presented in sections 2.3.5 and 2.3.6.

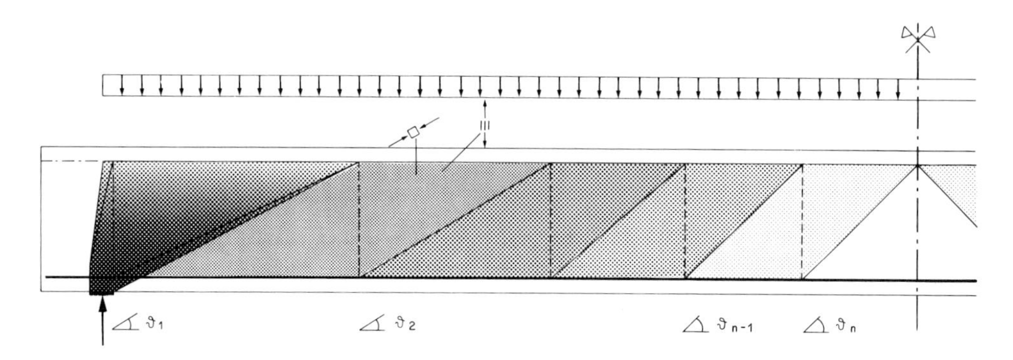

Fig. 2.30 Stress field with variable inclination of compression field.

2.3.3 General Case, Practical Design

The beam shown in Fig. 2.31 carrying a general loading at ultimate limit state is considered. The internal lever arm results from the assumption that the resultants of the compression and tension flanges act at the center of gravity of the flanges.

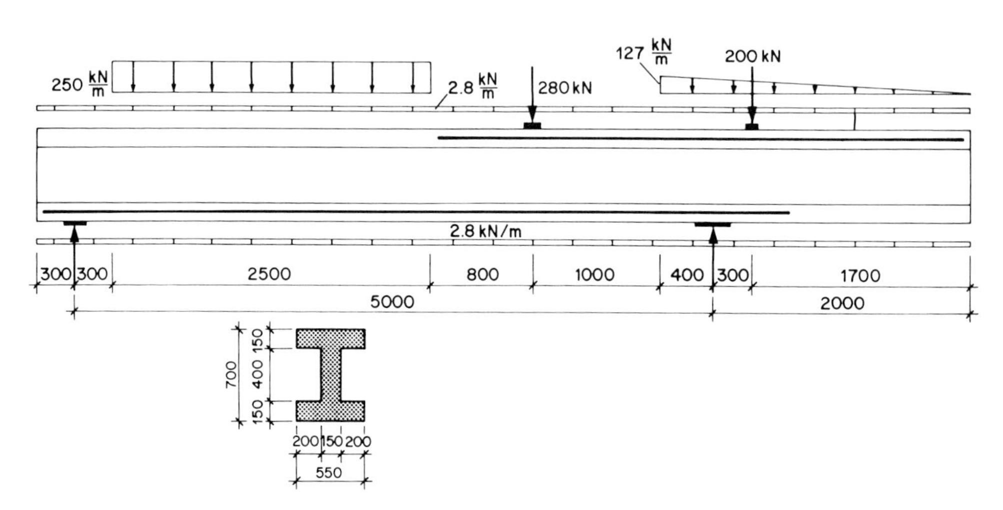

Fig. 2.31: Beam and loading

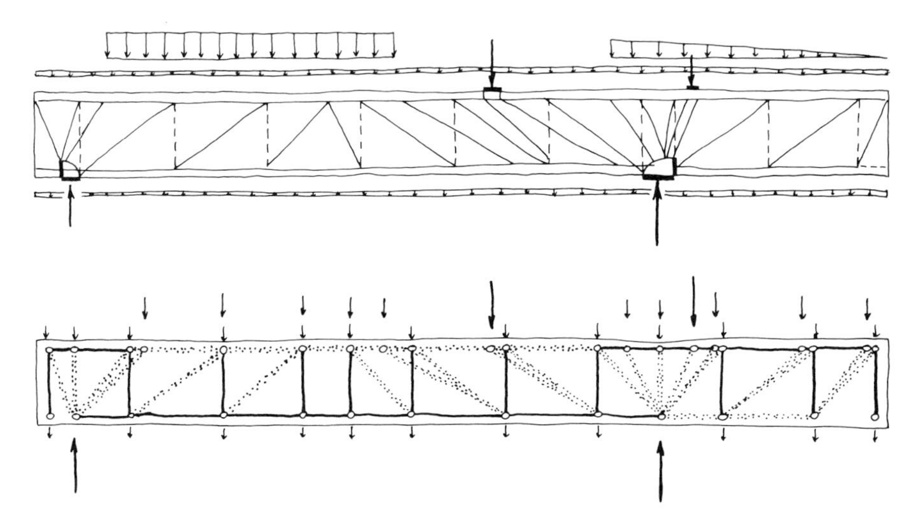

Fig. 2.32: Hand sketch of the stress field and the resultants

The stress field can be developed qualitatively by adopting elements of already known solutions (Fig. 2.32). The following points are taken into account:

- At the point of zero shear the inclination of the diagonal compression field changes direction with respect to the vertical. In the previous examples the zero point is located at midspan due to symmetry with respect to geometry and loading.

- The distance to the support of the concentrated load acting on the cantilever is sufficiently small to assume direct support. The concentrated load acting within the supported span, on the other hand, has to be supported indirectly. The stress field beneath this load is fanned out in such a way that a constant stirrup reinforcement results as shown in Fig. 2.33.

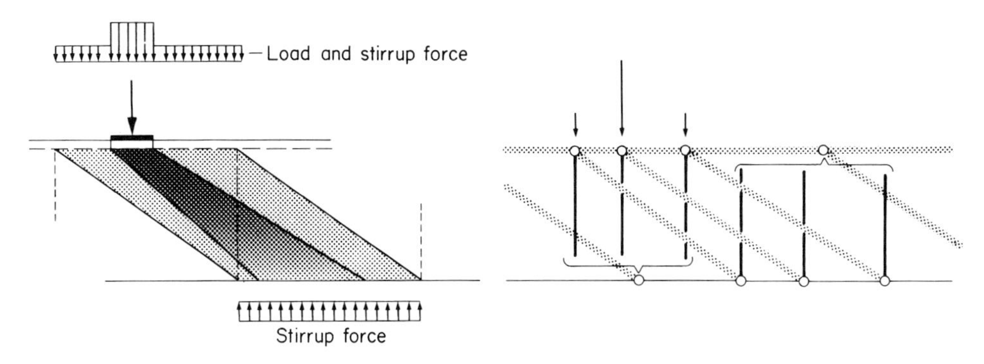

Fig. 2.33: Introduction of a concentrated load.

- A similar solution can be applied in the region of the cantilever, so that even with a variable load a constant stirrup reinforcement results sectionwise.

- The dead load acts with one half at the top and one half at the bottom of the beam.

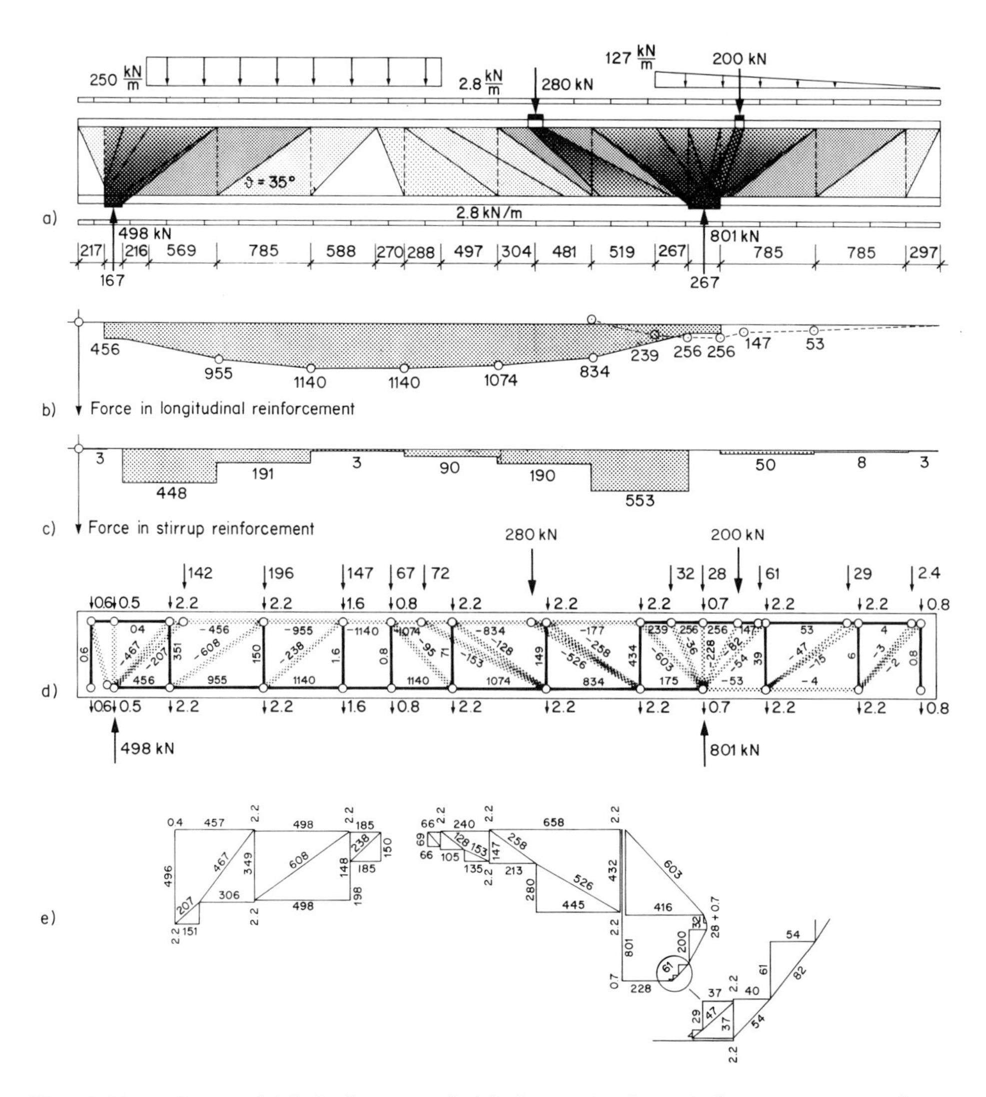

Fig. 2.34: *Integral Method*, stress field, forces in the reinforcement, resultants and Cremona diagram

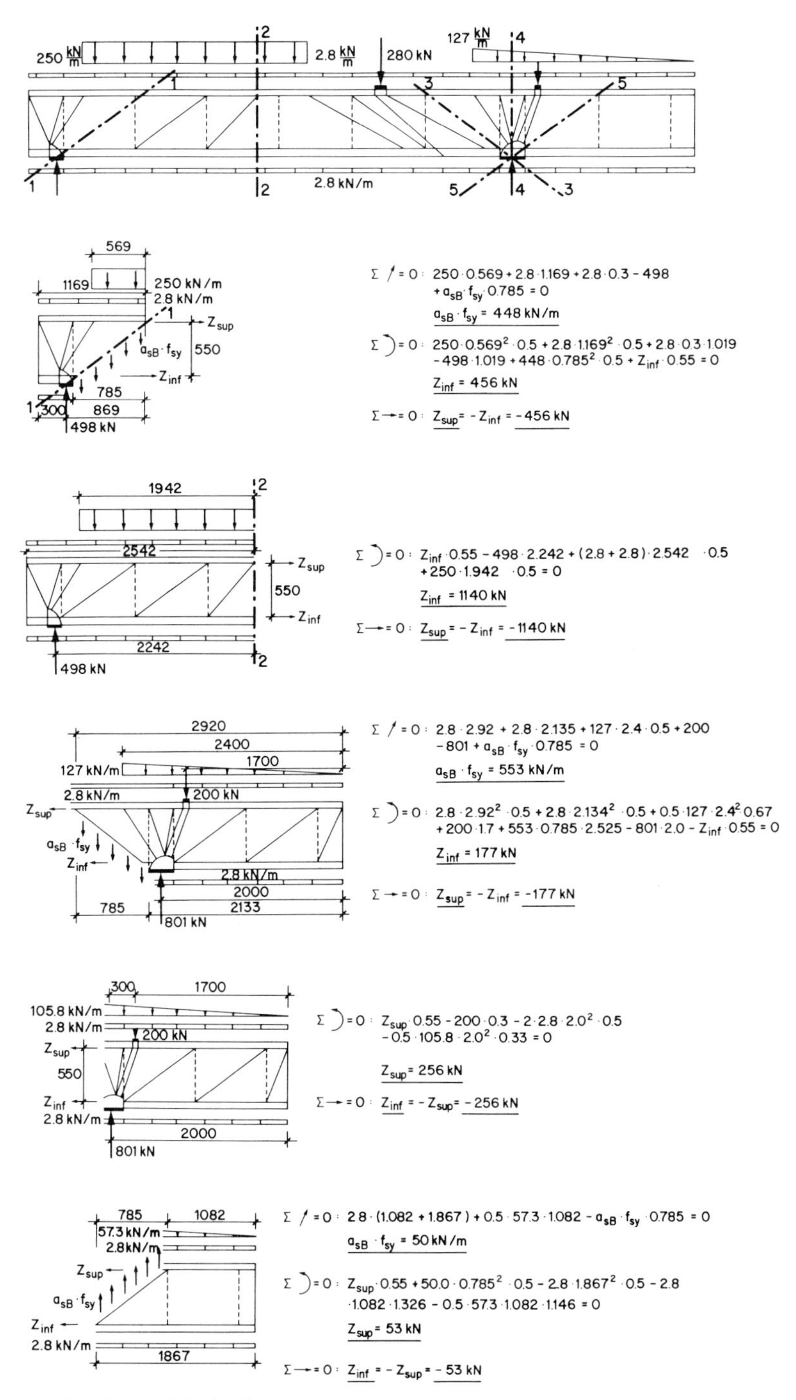

Fig. 2.35: *Sectional Method*

The final design of a beam can be carried out basically by one of two methods. The first method, called the *Integral Method* (Fig. 2.34), involves developing the complete stress field quantitatively.

As the examples dealt with above show, the resultants of the stress fields form a type of truss system. The stressing of the elements in the stress field results from the forces in the members of the truss, which may be determined successively by considering equilibrium at each joint.

In the design of simple structures it is not necessary to determine all the forces and stresses. The second method, called the *Sectional Method*, involves determining the internal stresses in critical regions using free-body diagrams with the aid of appropriate sections through the stress fields. A safe application of this method is only guaranteed if the variation of the complete stress field is known qualitatively.

Fig. 2.35 shows a possible application of this method. With the help of the sectional forces direct design equations may be derived to calculate the individual design values (Fig. 2.36):

$$\sum V=0:\quad V_A-(a_{sB}\cdot f_{sy}+q_{sup})\cdot z\cdot\cot\vartheta=0$$

$$a_{sB}=\frac{V_A}{f_{sy}\cdot z}\cdot\tan\vartheta-\frac{q_{sup}}{f_{sy}}$$

$$\sum M=0:\quad -M_A+Z_{infA}\cdot z-(a_{sB}\cdot f_{sy}+q_{sup})\cdot\frac{(z\cdot\cot\vartheta)^2}{2}=0$$

$$Z_{infA}=\frac{M_A}{z}+\frac{V_A}{2}\cdot\cot\vartheta$$

$$\sigma_{cA}=\frac{V_A}{\sin\vartheta}\cdot\frac{1}{z\cdot\cos\vartheta\cdot t}$$

$$\sum V=0:\quad V_B-(a_{sB}\cdot f_{sy}-q_{inf})\cdot z\cdot\cot\vartheta=0$$

$$a_{sB}=\frac{V_B}{f_{sy}\cdot z}\cdot\tan\vartheta+\frac{q_{inf}}{f_{sy}}$$

$$\sum M=0:\quad M_B+Z_{supB}\cdot z-(a_{sB}\cdot f_{sy}-q_{inf})\cdot\frac{(z\cdot\cot\vartheta)^2}{2}=0$$

$$Z_{supB}=-\frac{M_B}{z}+\frac{V_B}{2}\cdot\cot\vartheta$$

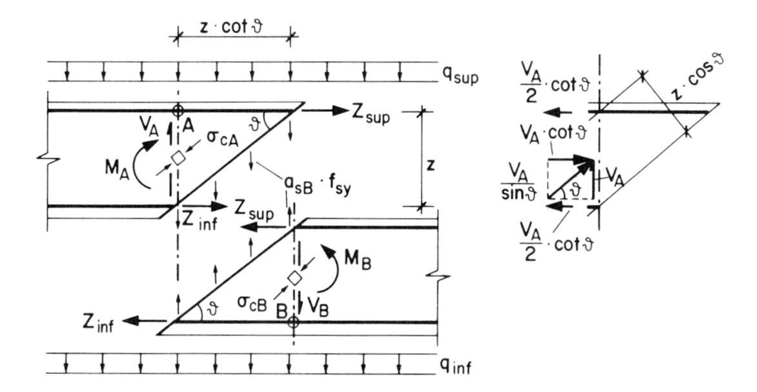

Fig. 2.36 Derivation of the internal forces from the sectional forces

2.3.4 Beams of Variable Depth

As an example a beam with the same span, loading and depth at midspan is taken as that treated in section 2.3.1. The depth of beam decreases linearly from midspan to the supports (Fig. 2.37a). The stress field can be developed in the same way as described in Fig. 2.28.

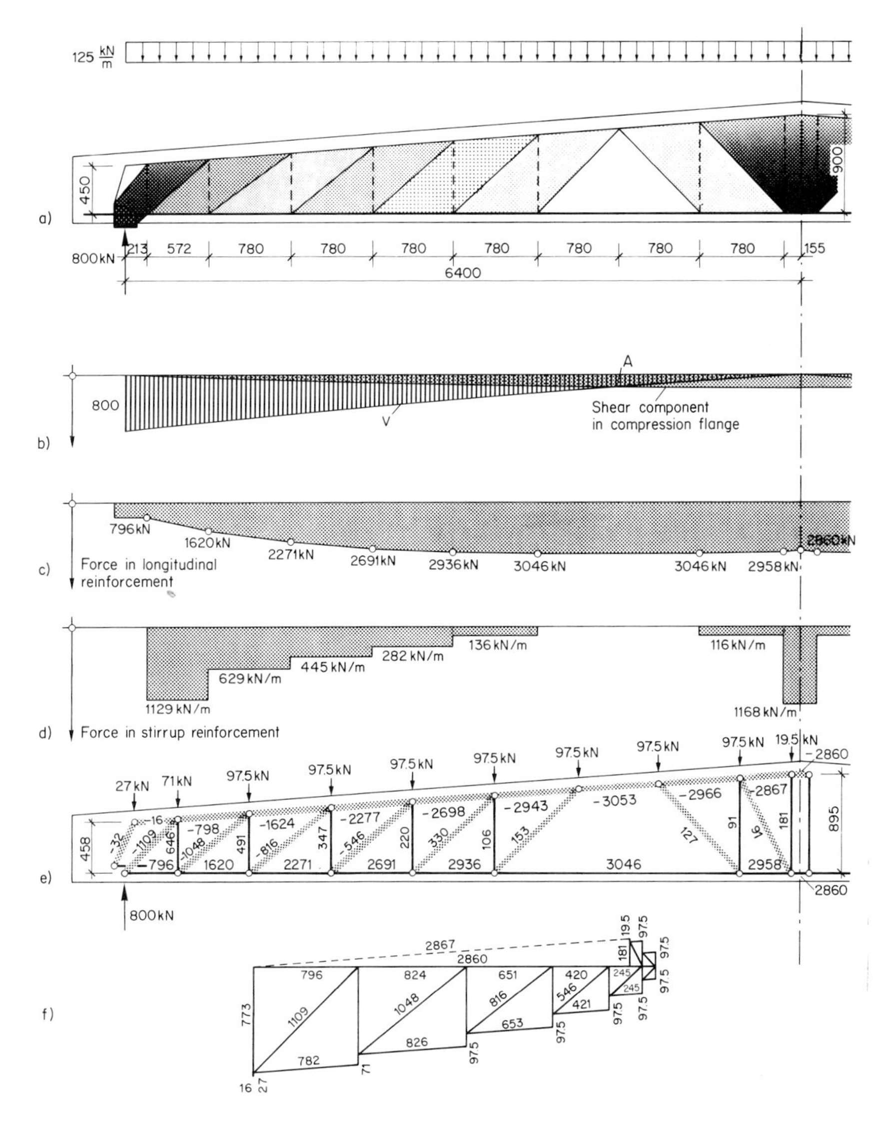

Fig. 2.37 : Stress field, forces in reinforcement, resultants and Cremona diagram

Thereby it has to be taken into account that the force in the inclined compression flange exhibits a vertical component which acts in the opposite direction to the shear force. This structural effect corresponds to a direct support. The variation of shear force is shown in Fig. 2.37b in comparison to that of the shear force component in the compression flange. All the shear force is transmitted through the compression flange at point *A*, so that the web remains unstressed. Between point *A* and the midspan the shear force component in the flange is greater than the acting shear force so that a negative shear force results in the web.

This means that in this region the loads are transmitted to the midspan, where they are directed upwards at the bend in the compression flange (deviation force) and transmitted to the supports. It should be noted that a variable statical height can result from sloping longitudinal reinforcement (e.g. tension member). Such a case will be dealt with in section 4.5.

2.3.5 Compression Flange

The compression flange of the beam of section 2.3.2 is investigated. The loading on this flange corresponds to the horizontal shear forces acting at the top of the web (Fig. 2.28). The spreading of these forces in the compression flange can be described by means of simple stress fields (Fig. 2.38). For the choice of the spreading angle the same considerations are valid as in the choice of the inclination of the compression zone in the web.

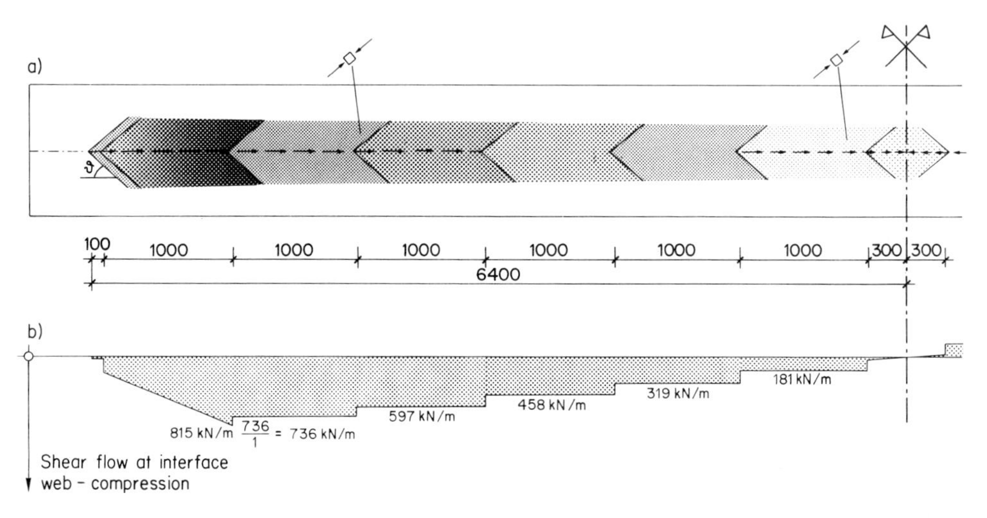

Fig. 2.38: Spreading of the shear forces in the compression flange

The inclined compression fields can be deviated by arranging transverse reinforcement in the flange. Fig. 2.39 shows this deviation in detail.

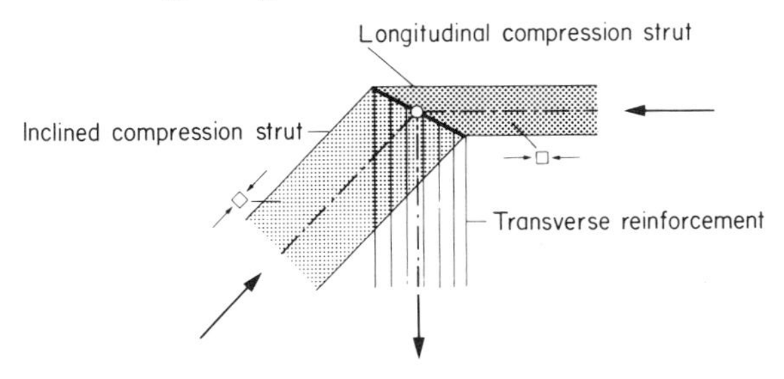

Fig. 2.39: Deviation of the inclined compression field in the flange

With this solution the deviation force due to the transverse reinforcement acts at the line of discontinuity between the two struts. The reinforcement force has to be anchored at or beyond this line of discontinuity.

It should be noted that this solution represents the general case of the deviation of a strut using reinforcement, in which all three fields have distinct dimensions. Special cases of this solution, in which either reinforcement or a strut act in a concentrated manner, are treated in section 2.2.1 (Fig. 2.8c) and in section 2.2.4 (Fig. 2.16b).

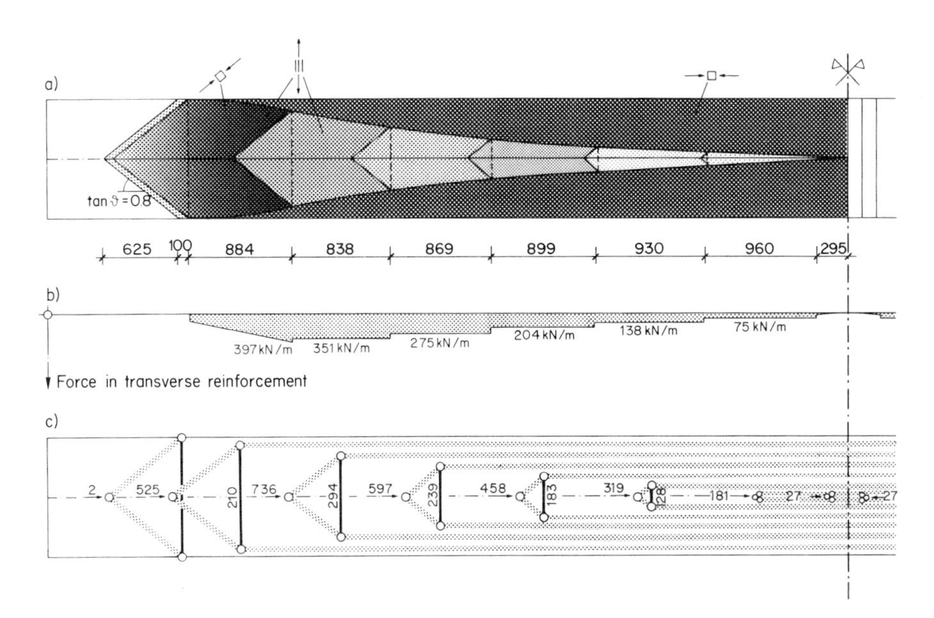

Fig. 2.40: Stress field, transverse reinforcement force and resultants

The widths of the inclined struts result from the dimensions of the struts in the web and from the spreading angle. The widths of the longitudinal compression struts result from the assumed distribution of the compression stresses at midspan. In Fig. 2.40 the complete stress field, including the transverse reinforcement force and resultants is shown, under the assumption that the distribution of the compression stresses at midspan is constant.

Strictly speaking this solution is valid only for an infinitesimally thin web. The consideration of the thickness of the web leads to stress fields with a smaller transverse reinforcement force.

2.3.6 Tension Flange

For the tension flange similar considerations apply as for the compression flange. Fig. 2.41 shows the stresses and their spreading in the flange.

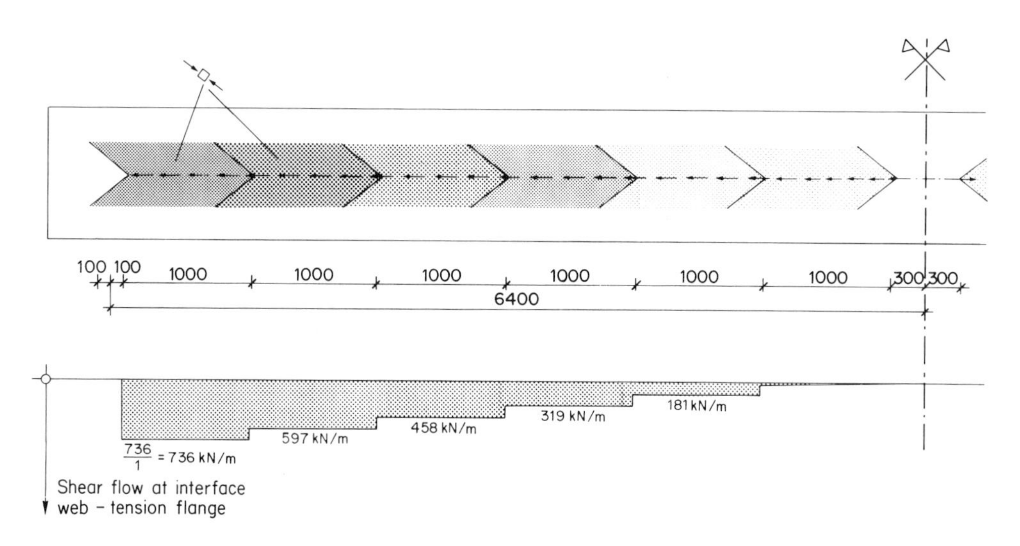

Fig. 2.41: Spreading of the shear forces in the tension flange

As for the compression flange the transverse force component of the diagonal compression field is resisted by transverse reinforcement, while the component in the longitudinal direction is resisted by longitudinal reinforcement as shown Fig. 2.42.

It should be noted that both the transverse and the longitudinal reinforcement have to be anchored at or behind the discontinuity line *AB*. Analogous to the compression flange the width of the inclined strut results from the size of the strut in the web and from the assumed spreading angle while the width of the individual tension fields result from the assumed distribution of the tension stresses at midspan.

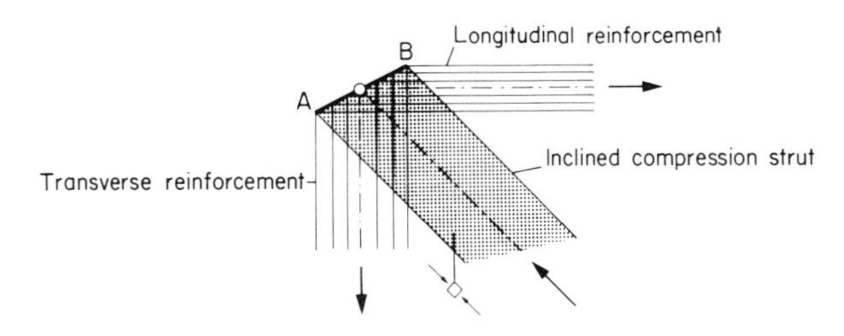

Fig. 2.42: Deviation of the inclined compression field in the flange

Besides the spreading forces shown in Fig. 2.41 there acts in addition, and in a concentrated way, the horizontal component of the force in the strut at the support (see Fig. 2.28). In this case this force is withstood by longitudinal reinforcement within the area of the web without spreading into the tension flange.

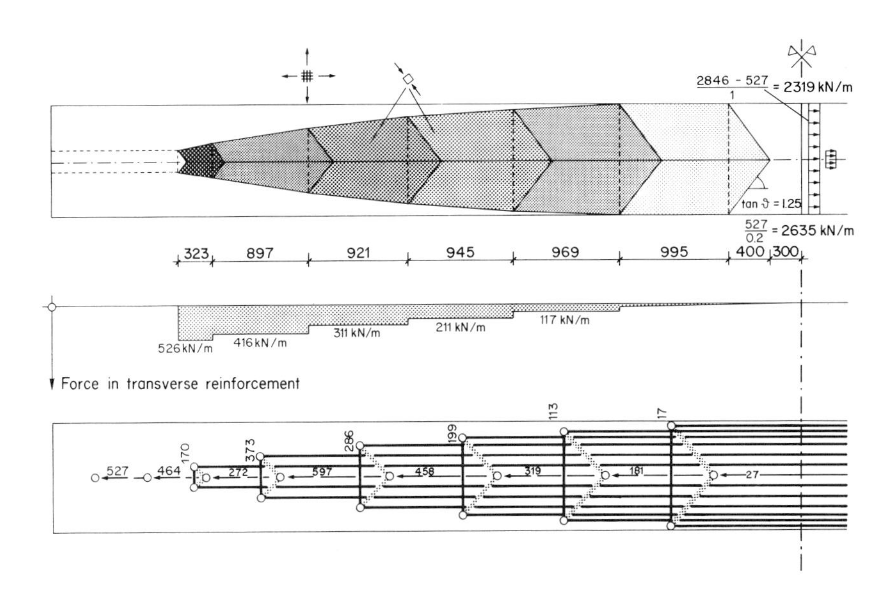

Fig. 2.43: Stress field, force in transverse reinforcement and resultants

For the other longitudinal forces a constant distribution in the transverse direction was chosen. Fig. 2.43 shows the complete stress field in the tension flange as well as the transverse reinforcement forces and all resultants.

In Fig. 2.44 a possible arrangement of the reinforcement for the web and the compression and tension flange is sketched.

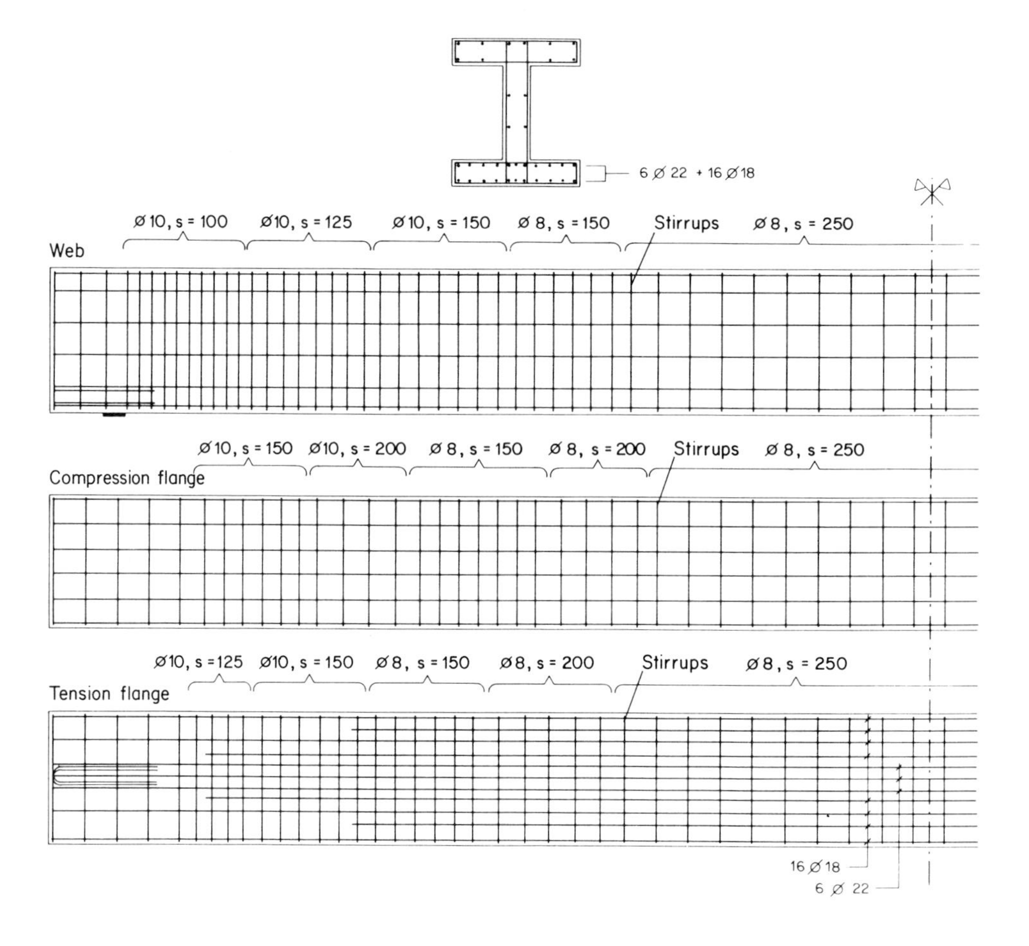

Fig. 2.44: Reinforcement sketch

2.4 MEMBERS SUBJECTED TO TORSION AND COMBINED ACTION

2.4.1 Introduction

In the examples treated in section 2.2 the forces are acting in the plane of the web of a member with symmetric cross-section. A general load can be carried if the cross-section consists of three plates which do not intersect along a common line. The load can be replaced by at least three statically equivalent forces, which act in the plane of the individual plates. The stress fields developed in section 2.2 can thus be applied correspondingly to the individual plates.

Members with a solid cross-section can be treated analogously by fitting into the section different suitable plates and neglecting the stresses in the rest of the material.

2.4.2 Warping Torsion (Open Cross-Sections)

As an application of the procedure described above the member shown in Fig. 2.45 will be discussed.

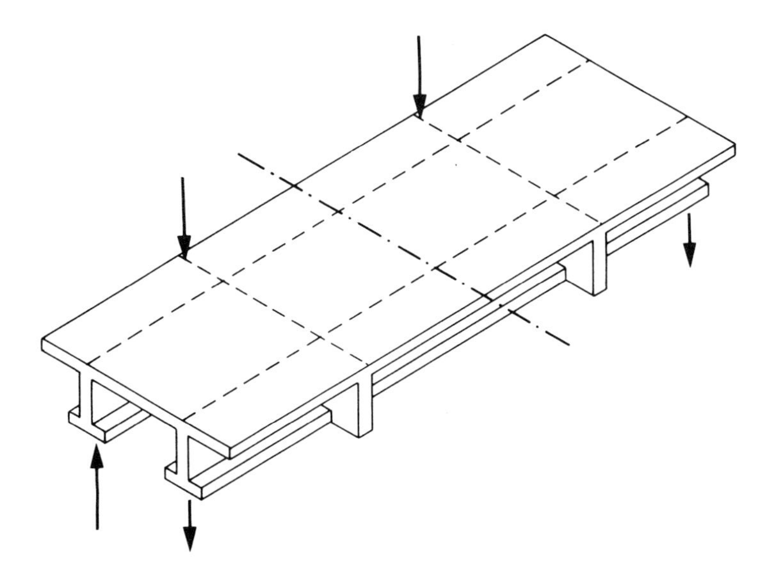

Fig. 2.45: Member with open cross-section

Since the line of action of the loads is parallel to the plane of the web, it is sufficient in this case to consider two plates. The introduction of the force into both webs results from cross-bending of the top compression flange or through a diaphragm as shown in Fig. 2.46.

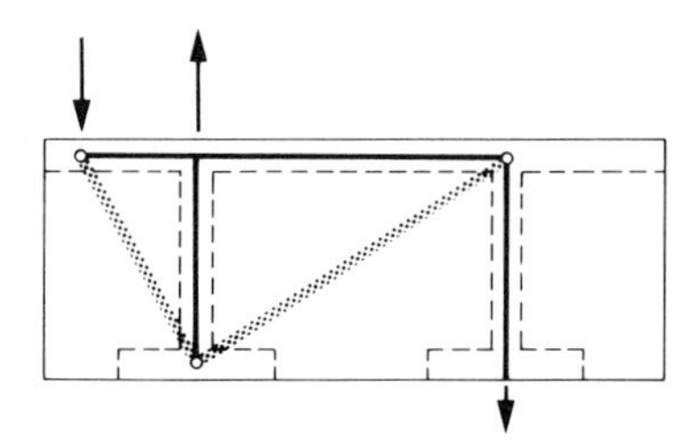

Fig. 2.46: Loading and stressing of the transverse diaphragm

The stress fields for both webs (see Fig. 2.47) and for the flanges can be adopted from section 2.3.

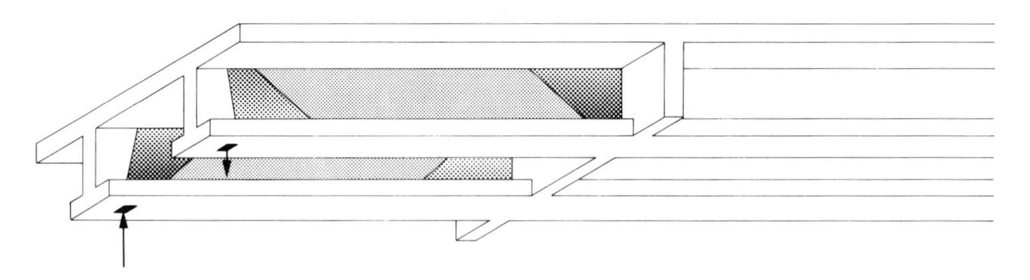

Fig. 2.47: Stress fields in webs

2.4.3 Circulatory Torsion

In the previous example the torsional moment is taken up by the two webs and the horizontal plates are only stressed by the resulting flange forces.

For a member with a closed cross-section (box section), in which all plates are subjected to resulting forces, it is possible in the case of pure torsion to choose the shear force in the individual plates in such a way that the shear flow along the interface of two flange plates is equal.

This is the case when the shear flow v (see Fig. 2.51) is the same in all plates. This structural action is illustrated in the following example (Fig. 2.48).

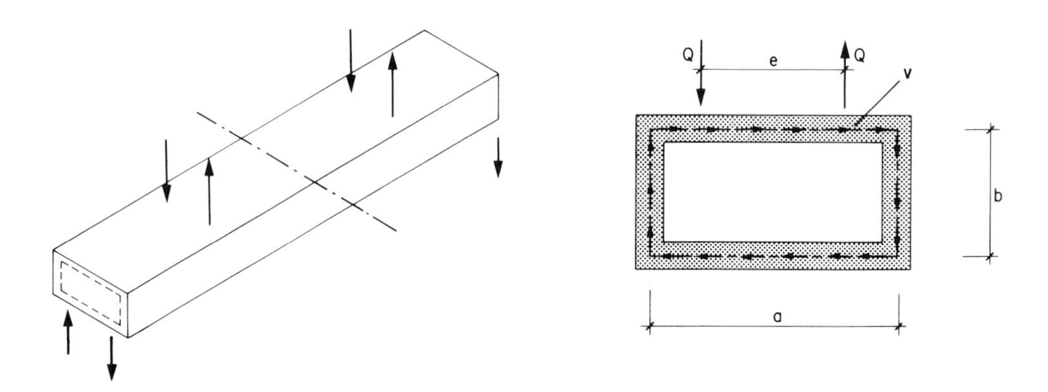

Fig. 2.48: Closed cross-section, system and loading

The shear forces in the plate are proportional to the heights of the plates. The introduction of the loads into the individual plates may result, as mentioned above, from either transverse bending of the plates or through a diaphragm (Fig. 2.49). The distribution of the internal forces in the diaphragm is investigated in section 2.9 by means of stress fields.

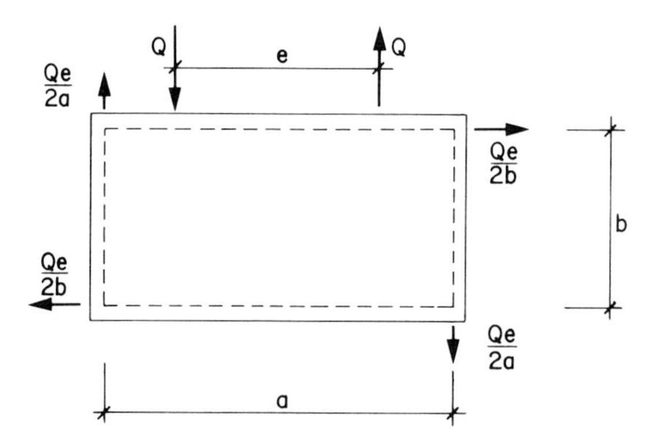

Fig. 2.49: Loading of the diaphragm

The forces acting on the individual plates as well as the corresponding stress fields are shown in Fig. 2.50.

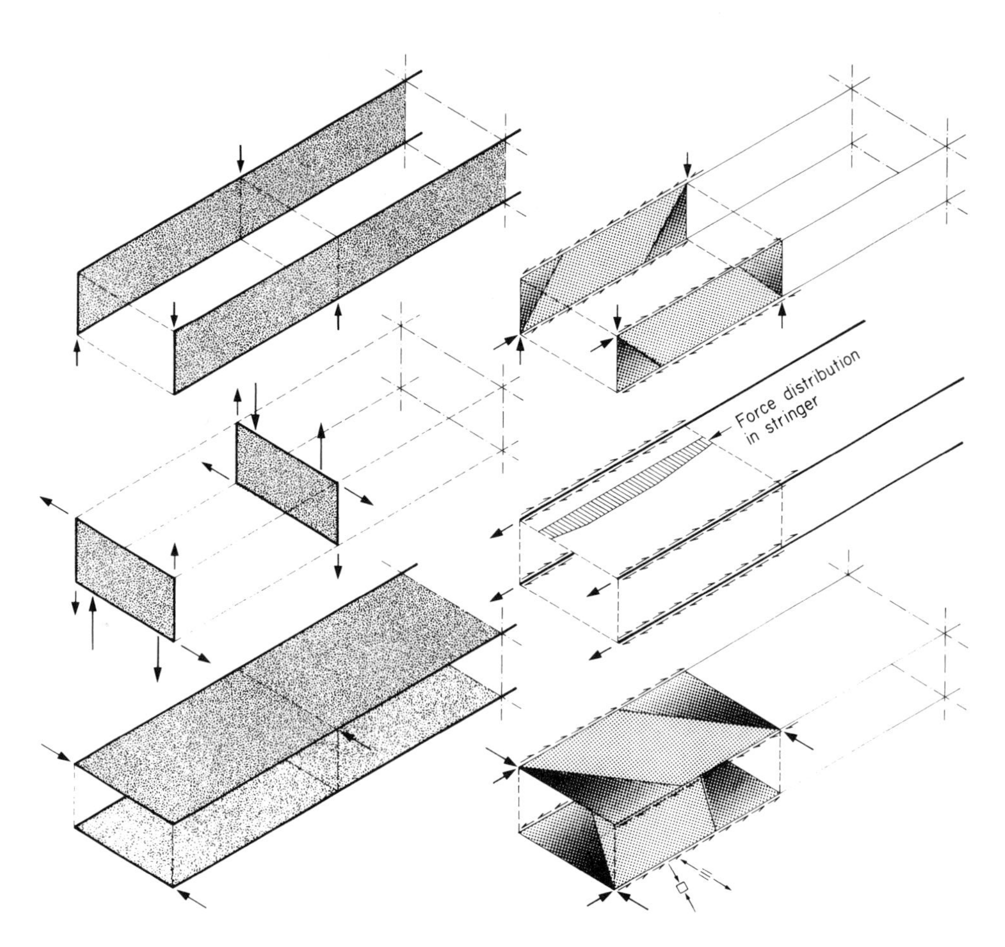

Fig. 2.50: Loading and stress fields in the flanges

The shear flow in two neighboring plate edges do compensate each other within the span region but not in the loading and support regions. Thus stringers loaded in tension result, which, if a cross-section is considered, maintain equilibrium with the longitudinal components of the inclined compression fields.

The longitudinal members may be designed separately. Thus the equations derived in Fig. 2.36 are also valid in this case. Such a derivation is given in Fig. 2.51 for a general cross-section.

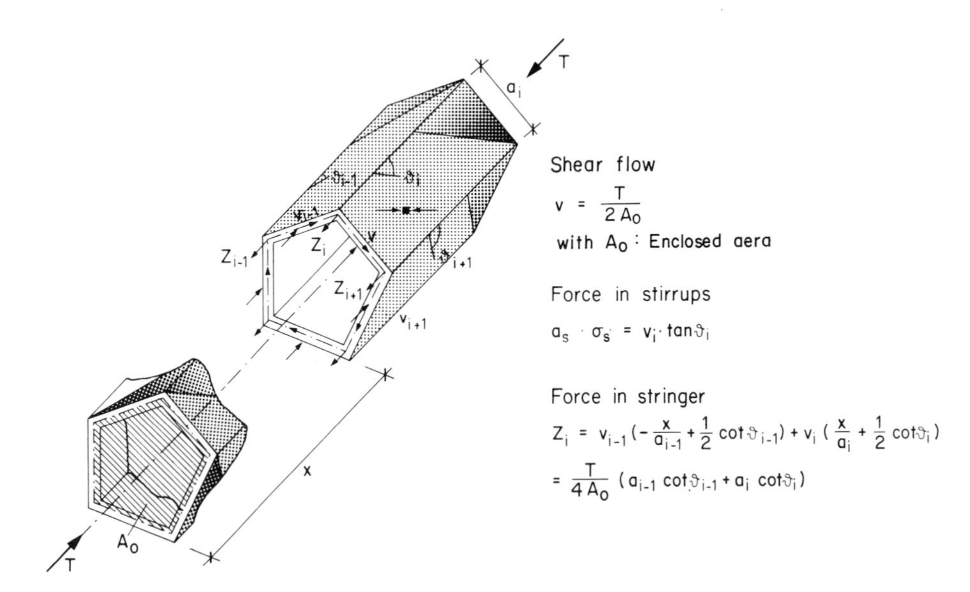

Fig. 2.51: Derivation of the design equations

In the remainder of this section a member subjected to a constant distributed torsional loading is investigated (Fig. 2.52). Since the loads have to be resolved into components which act in the planes of the plates, and as diaphragms are only present in the support regions, the longitudinal plates are also subjected to transverse bending.

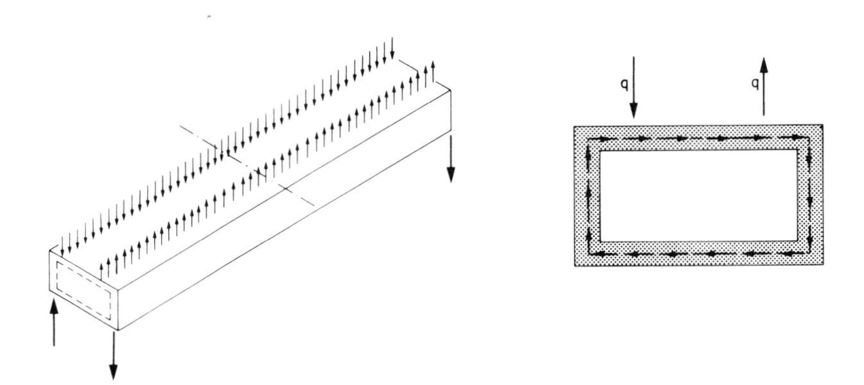

Fig. 2.52: Closed cross-section, system and loading

The forces acting on the individual plates together with the corresponding stress fields are shown in Fig. 2.53.

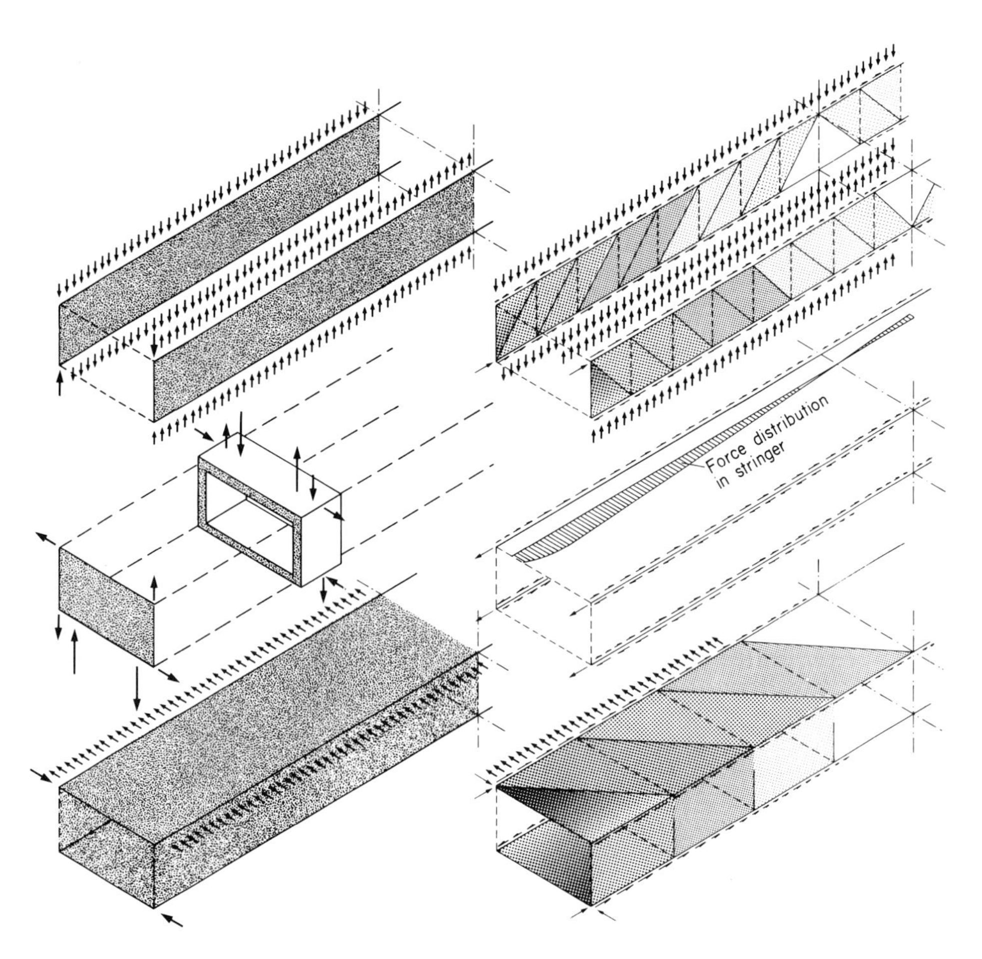

Fig. 2.53: Loading and stress fields in the plates

The plates are also subjected to transverse bending action. It is shown in chapter 5 how the combined action of transverse bending and shear in the individual plates has little influence on the stress fields.

2.4.4 Circulatory Torsion Combined with Bending and Shear

In practice, torsional loading is accompanied by bending and shear. A beam loaded under these conditions with the same geometry as the one treated in section 2.4.3 is shown in Fig. 2.54.

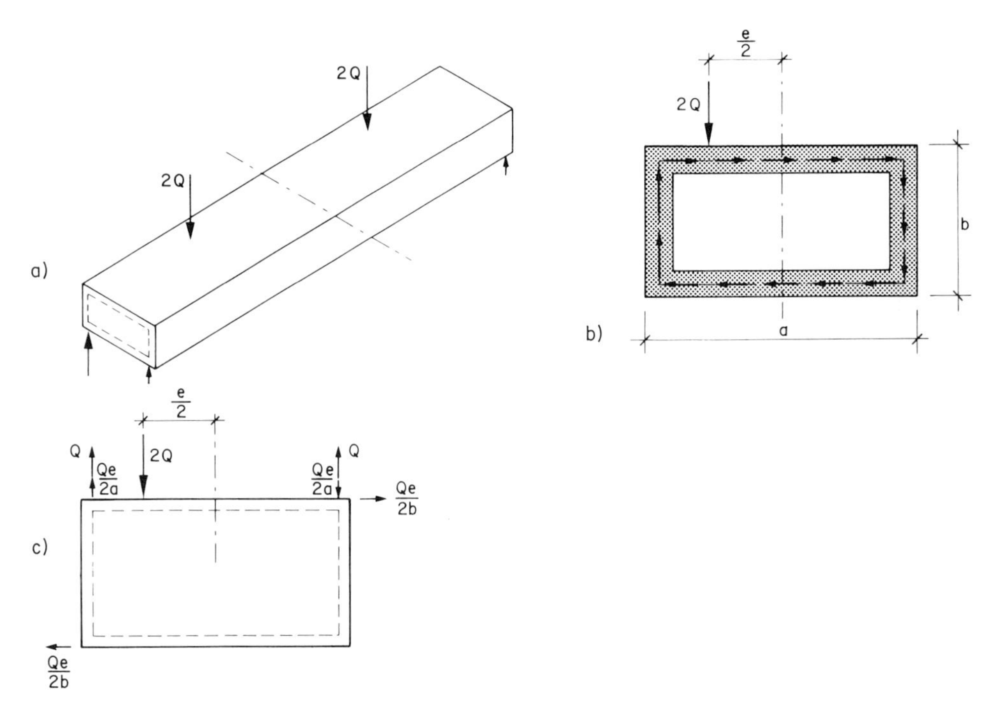

Fig. 2.54: Beam with box section under combined loading

The loading of the diaphragms and the webs (Fig. 2.54c) can be determined by superimposing the components due to torsion, bending and shear. Fig. 2.55a shows the loading of the webs. This results in known stress fields (Fig. 2.55b). The flange plates are acted upon by shear flows at the horizontal edges of the webs and by horizontal forces acting in the transverse plates. Fig. 2.55c shows associated stress fields, which were derived from those of a beam subjected to pure torsion, whereby the inclined compression fields in the middle region exhibit a double deviational change producing tension and compression forces in the longitudinal direction. Thus similar stress fields are formed in these regions to those obtained for the simple compression flanges (section 2.3).

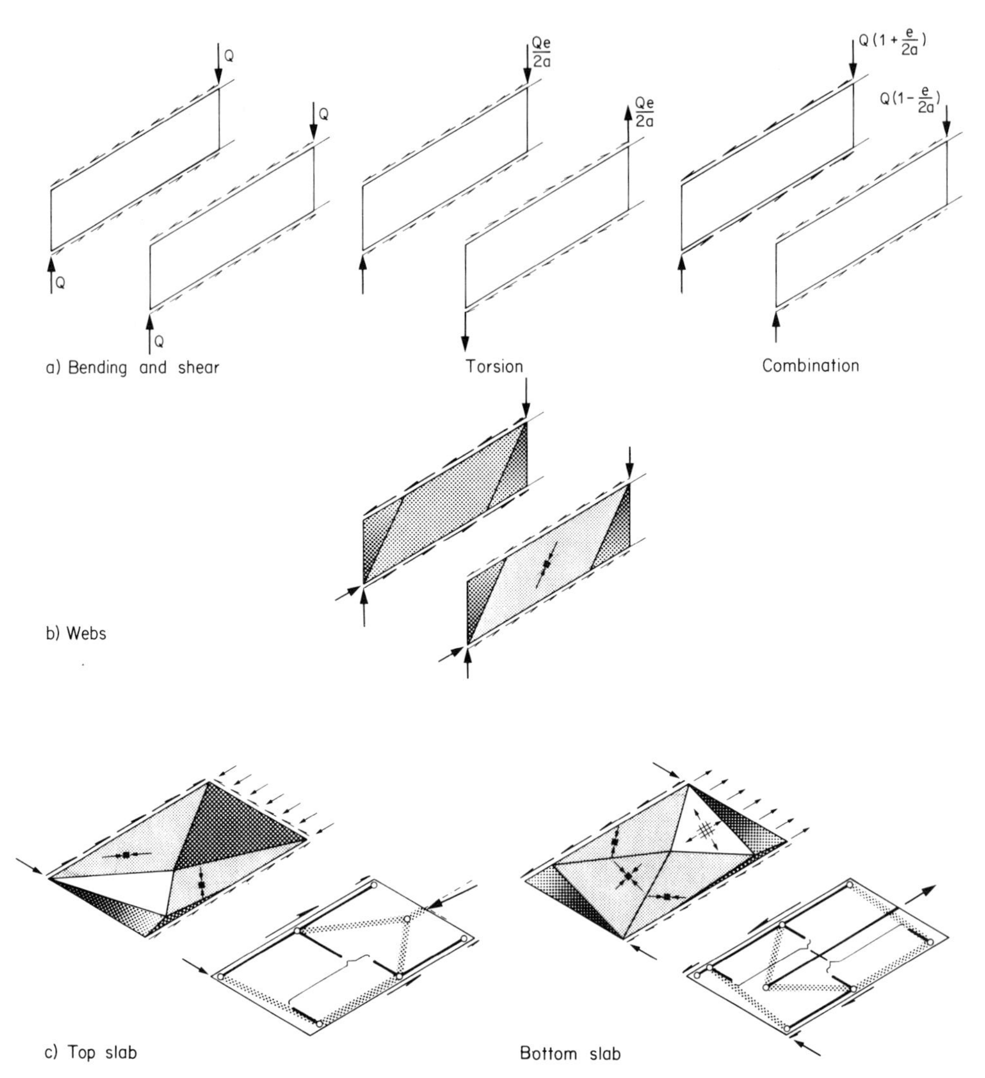

Fig. 2.55: Loading of the webs and the compression flanges

It is necessary to remark here also that for practical design purposes the entire stress field does not have to be determined quantitatively. The two important design quantities, namely the transverse reinforcement and the stress in the sloping compression fields, can be found from simple equilibrium considerations with the shear flow of the edge most highly stressed and with an assumed inclination of the compression field. From the condition that the amount of transverse reinforcement at both edges must be equal, the slope of the inclined compression fields is given at the lesser stressed edge.

2.5 BRACKETS

Brackets are generally characterized by a very small slenderness, so that a direct transfer of load to the support is possible. In Fig. 2.56 solutions are given for a bracket under both a concentrated load and a distributed load.

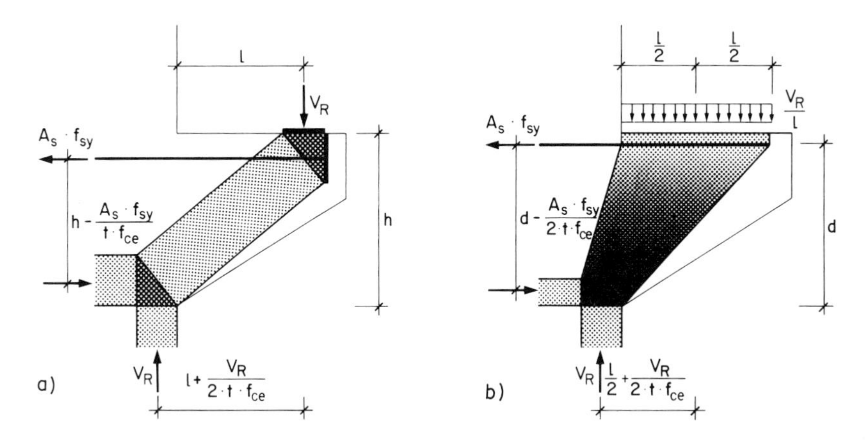

Fig. 2.56: Stress fields for brackets

Here again, for practical design purposes, it suffices to have a qualitative idea of the stress fields. The size and the position of the reinforcement may be found by means of simple equilibrium considerations. The reinforcement must be fully anchored either behind the position of load introduction (Fig. 2.56a) or in the region of load introduction by means of bond stresses (Fig. 2.56b).

2.6 COUPLING BEAMS

Coupling beams are subjected to forces and moments at both ends (Fig. 2.57).

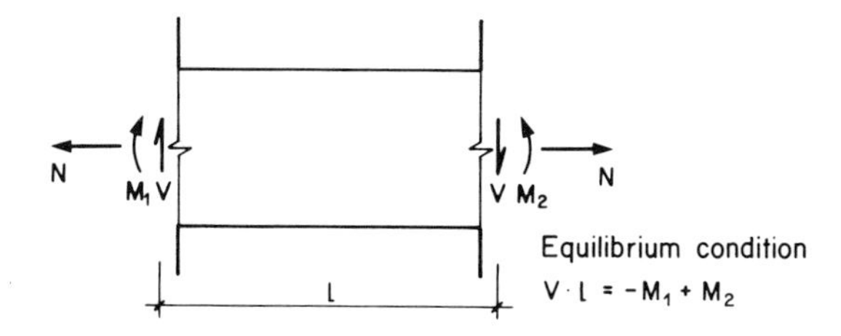

Fig. 2.57: Coupling beams, system and loading

The special case with $N = 0$ and $M_1 = 0$ corresponds to the examples dealt with in section 2.2. In the case of low slenderness of the member the shear force can be carried directly (Fig. 2.58a). The general case may be easily developed by introducing, in addition, tension and compression forces (Fig. 2.58c).

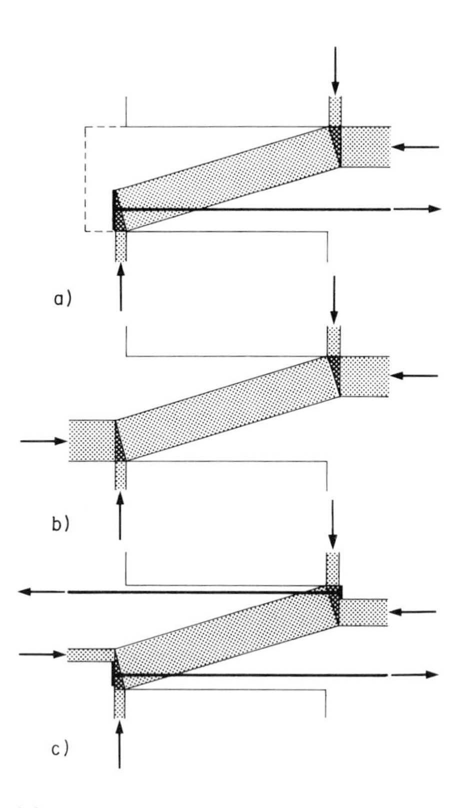

Fig. 2.58: Stress fields

As shown in Fig. 2.59a these fields may be simply constructed geometrically. Since the inclined strut is rectangular, points B and D lie on a circle through A and C. It follows that the shear force can be determined analytically as a function of the horizontal forces (Fig. 2.59b).

For very slender beams a strongly inclined diagonal strut results. If there is no longitudinal reinforcement loaded in tension present (Fig. 2.58b), then such a field can develop. If, on the other hand, the diagonal runs over a large stretch in the vicinity of horizontal tensile reinforcement (Figs. 2.58a and c), then the force transmission in the concrete is problematical (see chapter 4).

In these cases the load has to be transmitted indirectly (Fig. 2.60a) as dealt with in section 2.2.4. For beams of medium slenderness ratio the two systems of load carrying can occur in combination (Fig. 2.60b).

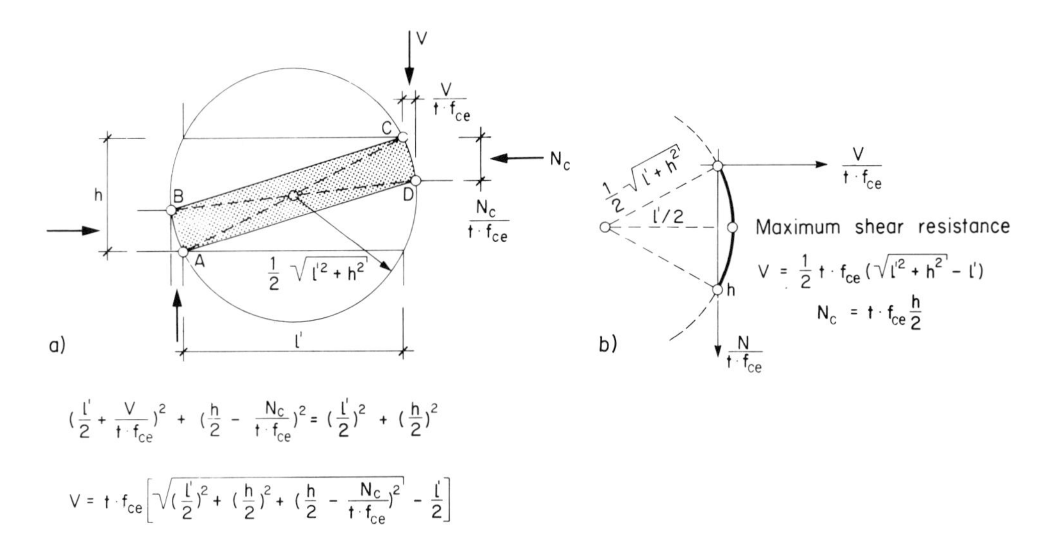

Fig. 2.59: Geometrical construction for the stress fields and *V-N* interaction

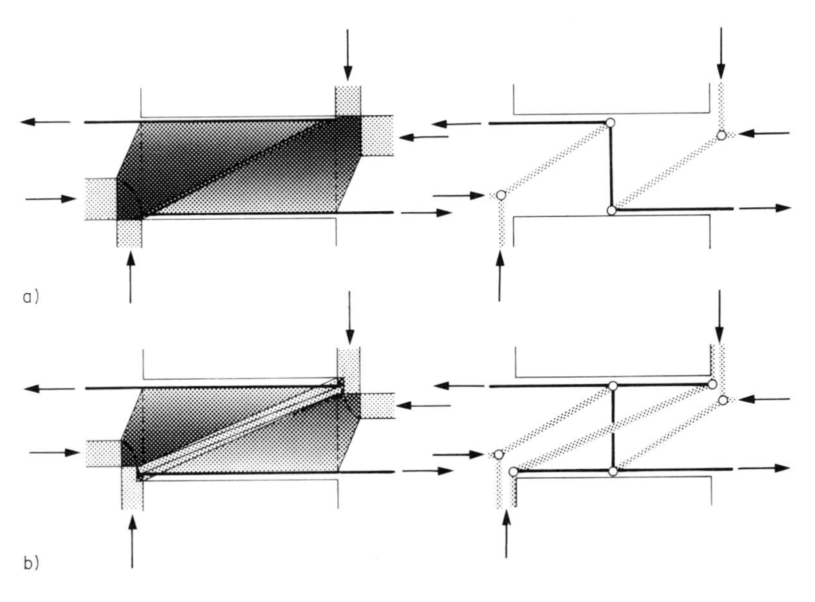

Fig. 2.60: Stress fields and position of the resultants

Short columns which are loaded by a normal force, shear force, and bending moment exhibit qualitatively the same structural action as for the connecting beams described above. Due to the high exploitation of the compression zone further considerations are required. Beam-columns are therefore treated in section 4.4.

2.7 JOINTS OF FRAMES

2.7.1 Corner Joint, Compression on the Inside

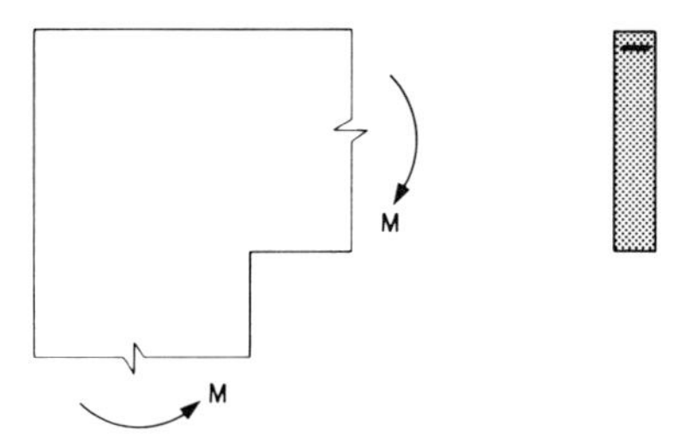

Fig. 2.61: Corner joint, geometry and loading

A corner joint loaded in bending shown in Fig. 2.61 is considered. Cases with additional shear and normal forces are dealt with at the end of this section. The dimensions of the compression strut as well as the reinforcement force in the individual beams can be obtained from simple static considerations (Fig. 2.62a). The horizontal compression struts meet in the region of the joint where they are in equilibrium with a compression diagonal (Fig. 2.62b).

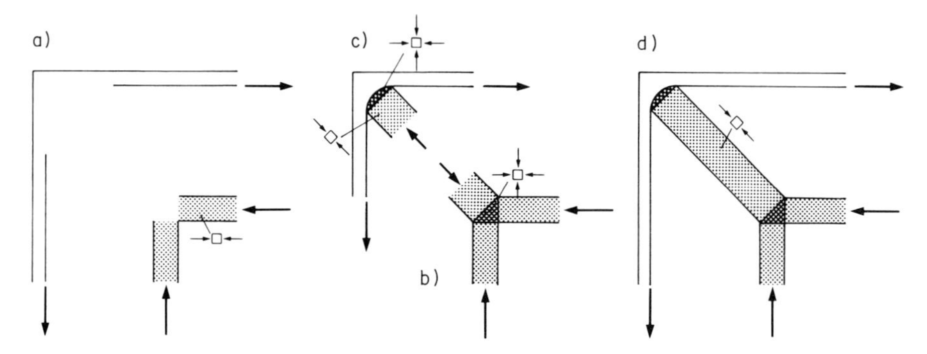

Fig. 2.62: Development of the stress field

Analogously the reinforcement force is deviated by this diagonal (Fig. 2.63c). As the magnitude of the compression and tension forces are identical, the magnitude of the forces at the ends of the diagonal are also identical.

The deviation of the reinforcement force is shown in detail in Fig. 2.63. Theoretically, the zone between the reinforcement and the compression diagonal is stressed biaxially because the reinforcement is bent into a circular form. For practical purposes, however, this is of no significance.

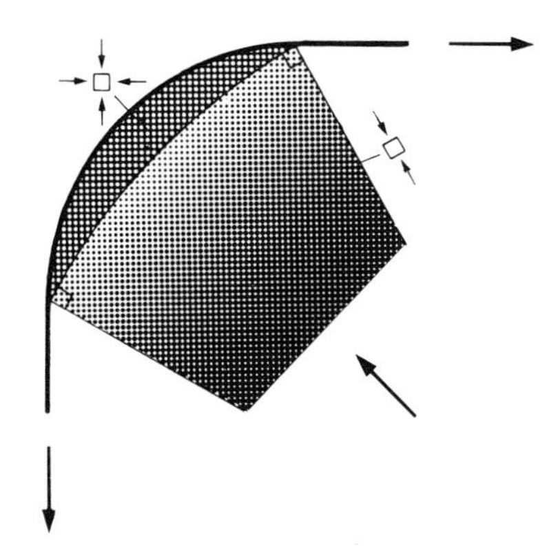

Fig. 2.63: Deviation of the reinforcement force

It should be observed that this derivation is not only valid for symmetrical cases. Fig. 2.64 shows the stress field for the corner joint, where the two joining members have different statical heights.

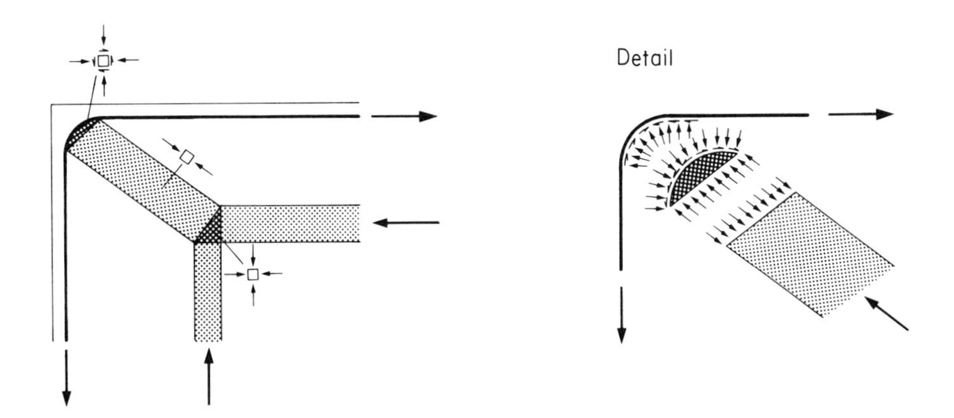

Fig. 2.64: Stress field for corner joints with legs of different statical heights

In this case the deviation of the reinforcement force is accompanied by bond stresses between the diagonal and the reinforcement. The equilibrium is only fulfilled for a specific shape of the reinforcement which is no longer circular. This can lead to problems under extreme conditions (small inclination of the strut and large radius of curvature of the reinforcement in comparison with the small statical height).

The stress field for the case of pure bending of a corner joint can be generalized by introducing additional normal and shear forces. Two such cases are shown qualitatively in Fig. 2.65.

It should be observed that also in these cases the reinforcement necessary from the statical point of view can be determined by the *Sectional Method.*

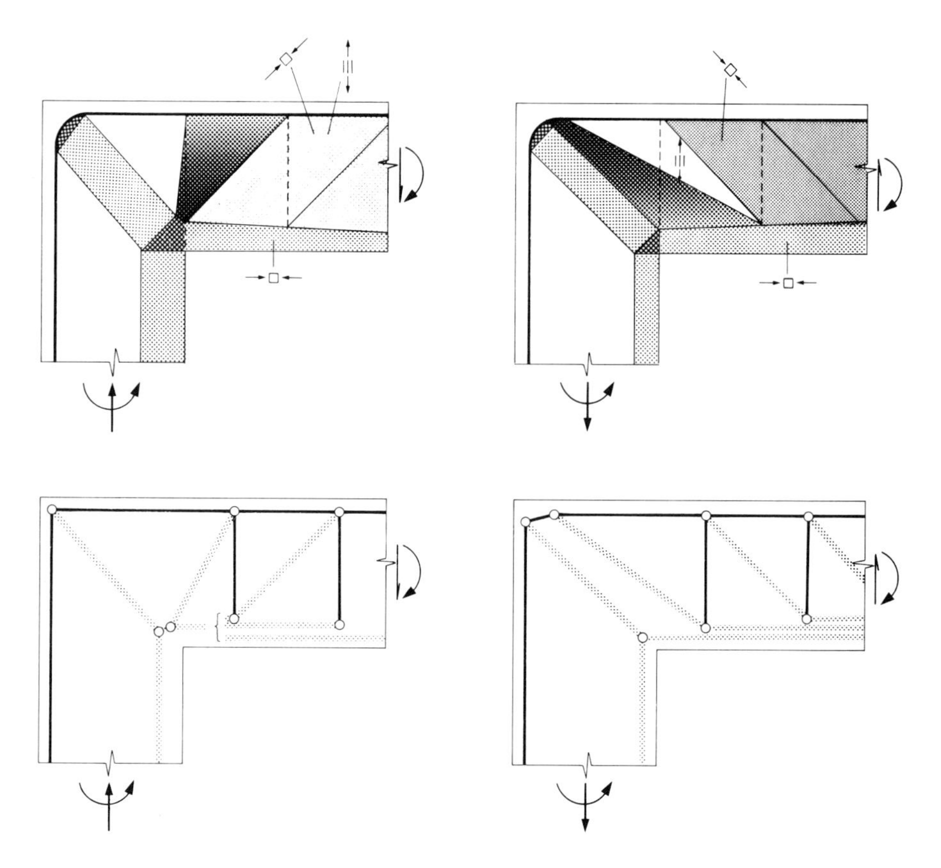

Fig. 2.65: Stress fields of corner joints under moment, normal and shear forces.

2.7.2 Corner Joint, Tension on the Inside

The stress field for corner joints stressed in such a manner can be taken from section 2.7.1. The tension members have to be replaced by struts and vice versa (Fig. 2.66a).

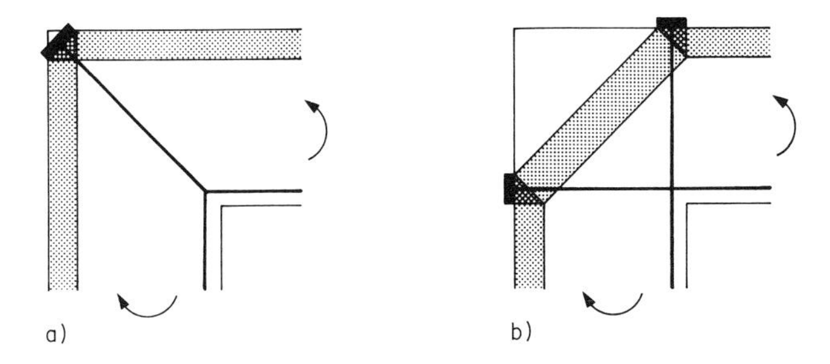

Fig. 2.66: a) simple deviation, b) double deviation of the compression strut

An alternative solution with a double deviation of the compression strut is shown in Fig. 2.66b. It should be noted that the deviation of the compression strut has to be carried right up to the surface of the concrete. This can only be achieved in detail by using external anchor plates. Experimental investigations have shown that all other solutions (without the aid of additional reinforcing bars) lead to unsatisfactory results due to the unsufficient anchoring conditions. The influence of the positioning of the reinforcement for specific problems is discussed in chapter 3. The problem is lessened if the strut does not have to be deviated completely to the outside edge of the joint. Fig. 2.67 shows such a solution in which there is an additional deviation of the strut by means of additional inclined reinforcement.

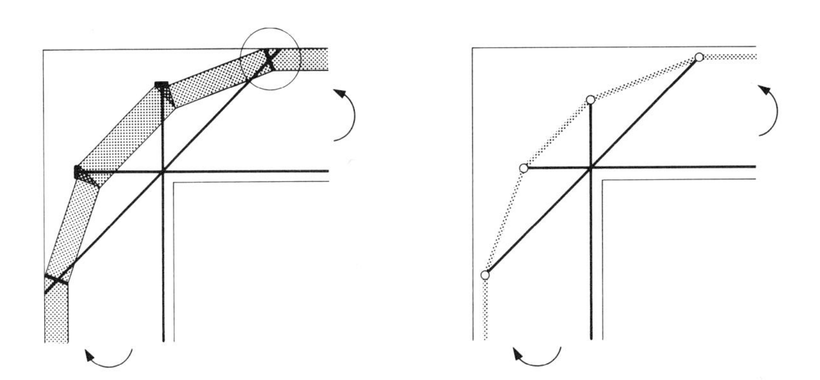

Fig. 2.67: Stress field and resultant (multiple strut deviation)

The deviation at region A is analogous to that of Fig. 2.39. A further possibility of avoiding the problems associated with a concentrated deviation is to arrange the stirrups in a distributed manner. In this way the strut can be deviated continuously (Fig. 2.68a). The forces in the stirrups have to be transferred to the main reinforcement (Fig. 2.68b). The final stress field and the corresponding resultants are shown in Figs. 2.68c and d.

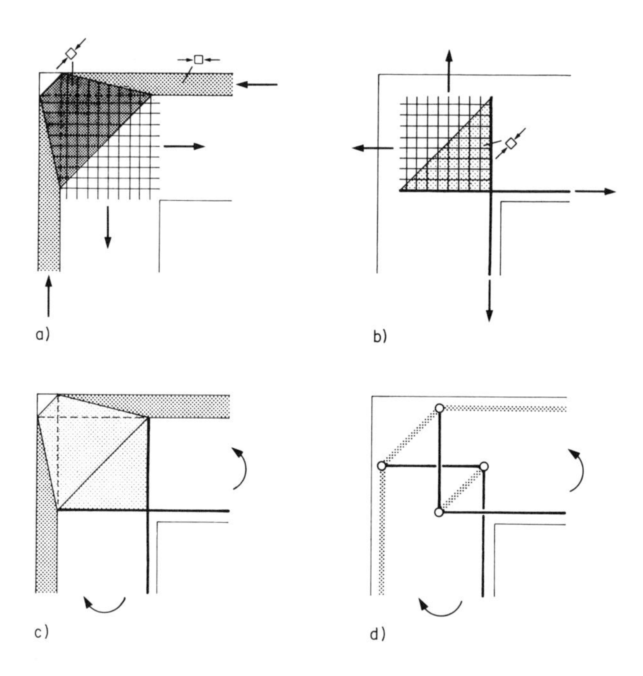

Fig. 2.68: Action of stirrups in corner joint

Fig. 2.69 shows a possible combination of stress fields according to Figs. 2.66b and 2.68.

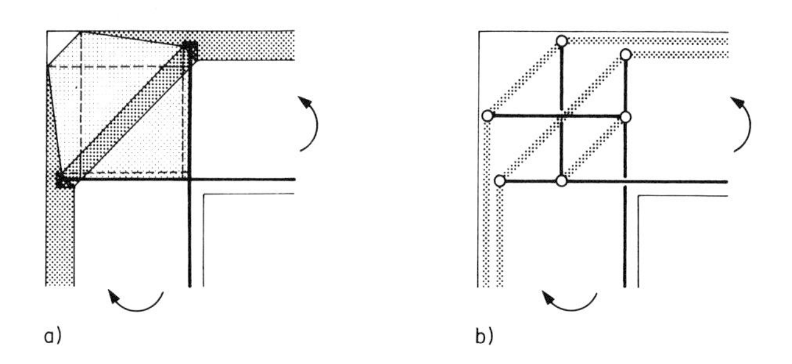

Fig. 2.69: Stress fields and resultants in corner joint

The change of the stress field of Fig. 2.67 due to the introduction of normal and shear forces is shown in Fig. 2.70. The reinforcement which corresponds to the stress fields of Figs. 2.66b, 2.67 and 2.68 is sketched in Fig. 2.71.

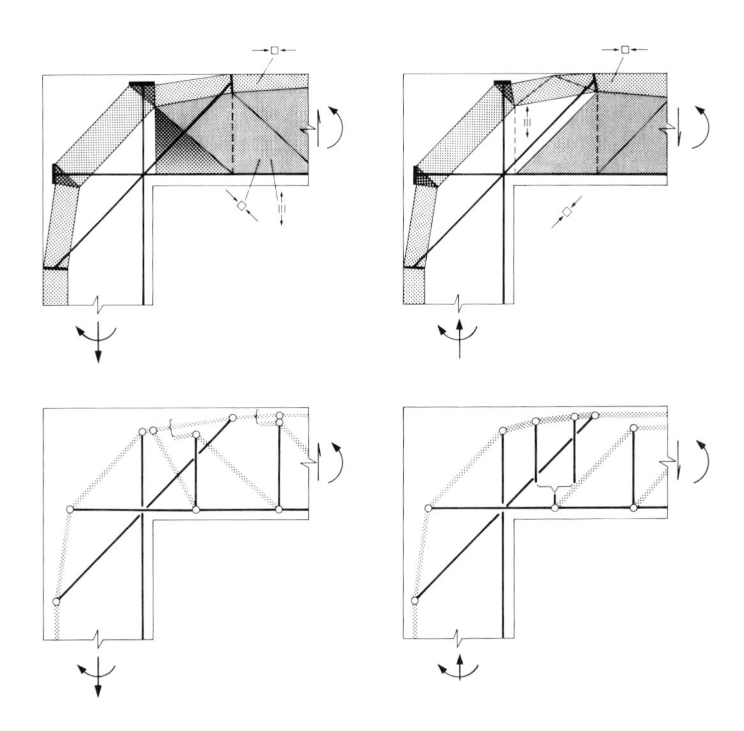

Fig. 2.70: Stress fields of corner joints under moment, normal and shear forces

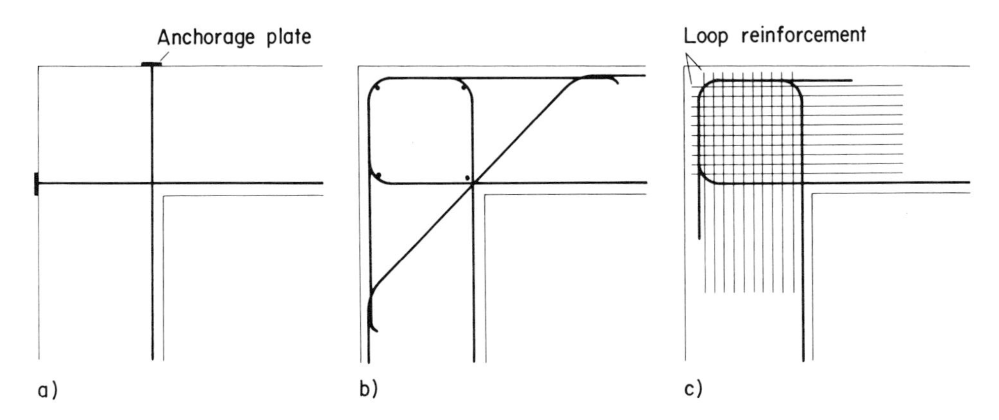

Fig. 2.71: Reinforcement sketches for corner joints

2.7.3 Joints of Frames with Three Connecting Beams

The discussion in this section is restricted to joints subjected to pure bending. Normal and shear forces influence the stress fields in a similar manner as in the case of the corner joints treated previously. The three possible types of loading are shown in Fig. 2.72.

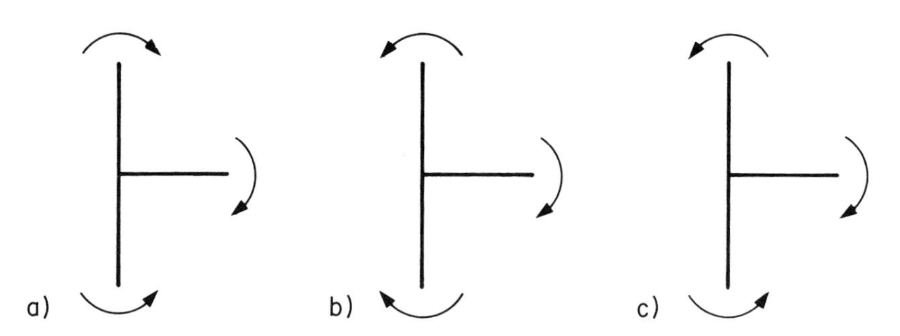

Fig. 2.72: Loading types of joints

A qualitative investigation of the internal forces with the aid of the resultants leads to the solution shown in Fig. 2.73.

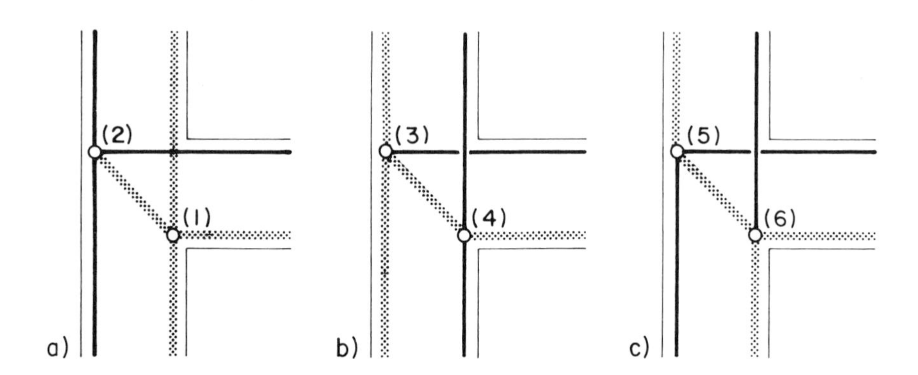

Fig. 2.73: Resultants

After fixing the dimensions of the strut the development of the stress field is restricted to the region of the joint where the struts meet. Possible solutions are shown in Fig. 2.74.

The details 1, 2 and 4 present no detailing problems to achieve equilibrium. For the details 3 and 5, however, the same anchoring difficulties arise as encountered for the joints of frames having tension on the inside dealt with in section 2.7.2.

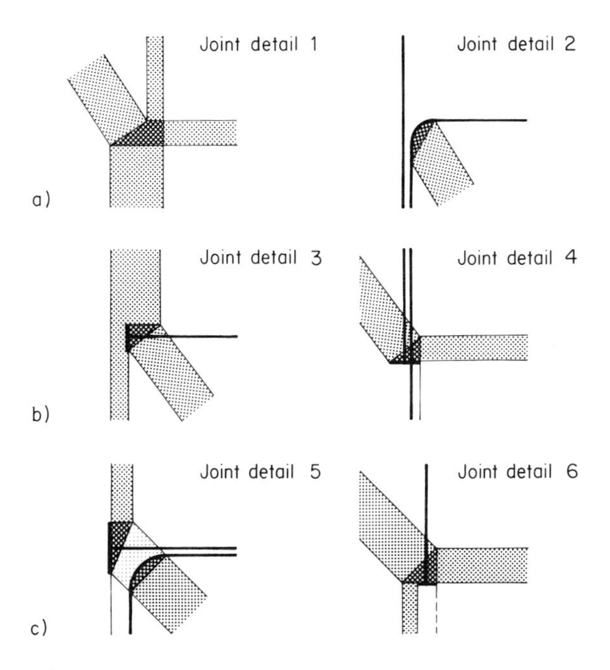

Fig. 2.74: Details

In the case of detail 5 one part of the diagonal force is balanced by the deviation force of the reinforcement. In the upper part of the diagonal the stress is therefore reduced and in the uppermost biaxially stressed region one of the principal stresses is hence less than f_{ce}. In detail 6 either the vertical reinforcement or the vertical strut (assuming that the stress intensity is f_{ce}) must be moved to the left, as otherwise an overloading of the biaxially stressed region would result. The complete stress fields are shown in Fig. 2.75.

For the joints (Figs. 2.75b and c) the problem of anchoring the horizontal reinforcing bars can be alleviated by introducing an additional inclined reinforcing bar or vertical stirrups as for joints with tension on the inside (Figs. 2.67 and 2.68). As shown by the stress field for the joint (Fig. 2.75c), the incomplete exploitation of the effective statical height in the lower beam has such a small influence on the required reinforcement that this effect can be neglected for practical design purposes.

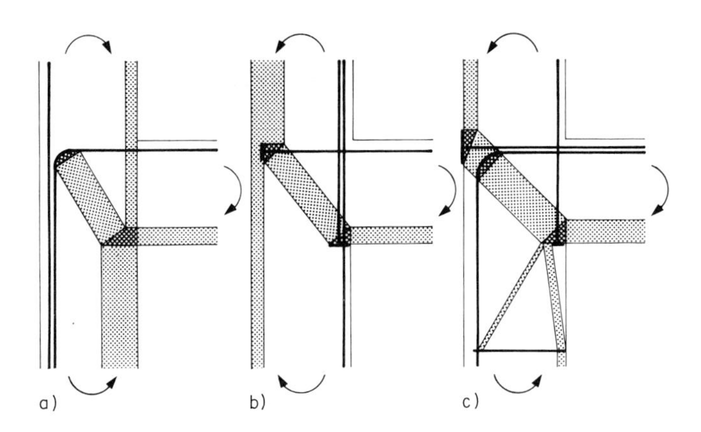

Fig. 2.75: Stress fields

2.7.4 Joints of Frames with Four Connecting Beams

For these joints the same considerations apply as for those with two or three connecting members. Fig. 2.76 shows the three possible types of loading and the resultants of the internal forces.

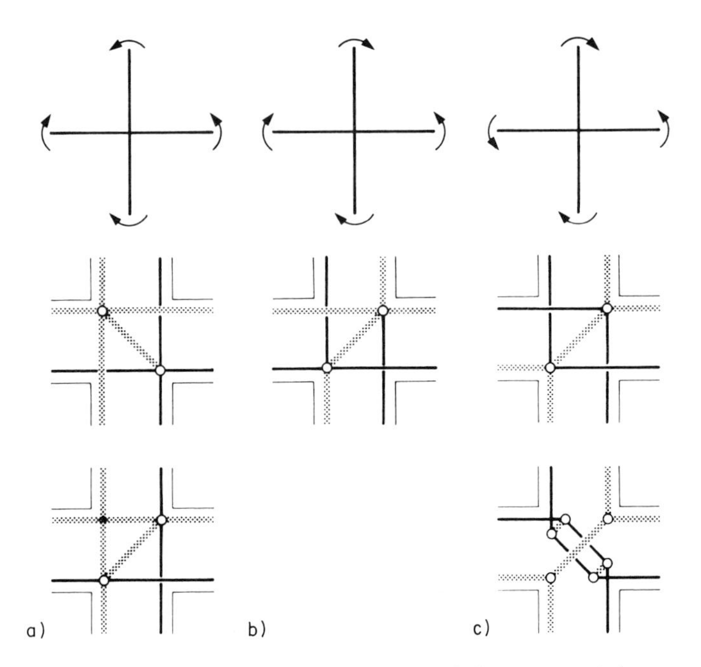

Fig. 2.76: Types of loading and resultants of the internal forces

In general, in the design of joints, the following procedure can be applied:

- Investigation of the distribution of the internal forces with the aid of the resultants.
- Determination of the required reinforcement cross-sections. The reinforcement forces usually result from those of the adjoining members.
- Investigation of the joint details with stress fields.
- Layout of the reinforcement.

2.8 BEAMS WITH SUDDEN CHANGES IN CROSS-SECTION

Fig. 2.77 shows two beam elements with a sudden change in depth subjected to pure bending.

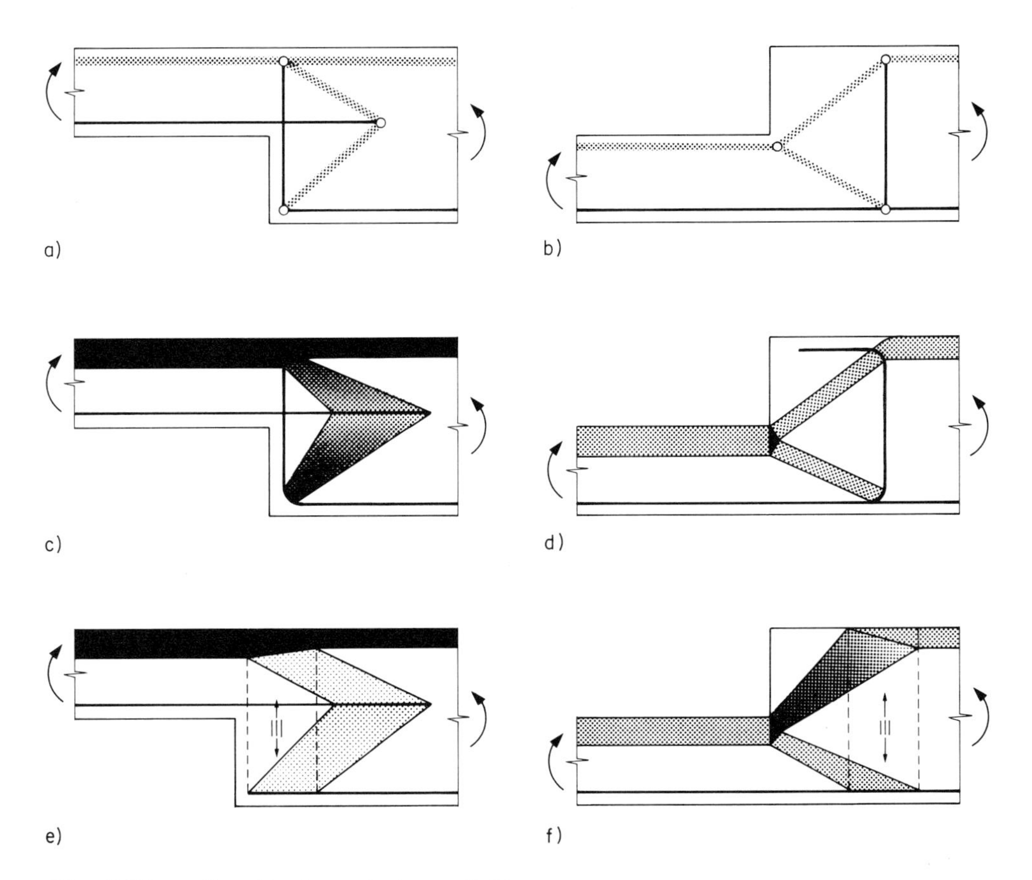

Fig. 2.77: Resultants and stress fields

The tension and compression forces outside of the zone of disturbance can be determined here too from simple bending considerations. The variation of the internal forces in the transition zone can be investigated with the aid of the resultants (Figs. 2.77a and b). The detailing of the reinforcement should be based, however, on the stress fields (Figs. 2.77c and d).

As is evident from the stress fields, above all, the reinforcement details in the region of the deviation of the longitudinal strut should be investigated with care. This problem can be alleviated by introducing stirrups allowing a distribution of the deviation force (Fig. 2.77e and f).

Fig. 2.78 shows solutions for an element which is stressed in addition by shear forces. Here as well the anchoring of the transverse reinforcement in certain regions can be problematic (Fig. 2.78a). A reasonable solution is achieved by introducing sloping reinforcement (Fig. 2.78b).

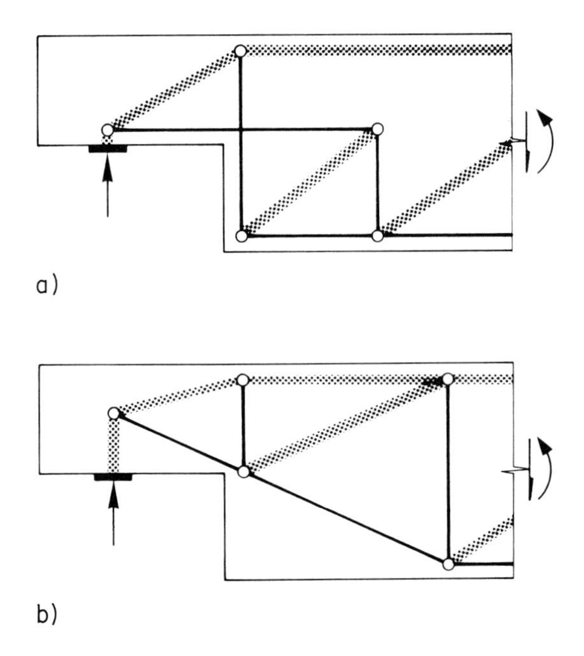

Fig. 2.78: Resultants in the region of a support

For beams with holes in the web the main concern is the transmission of the shear force. The region of the hole is bridged by two connecting beams, so that the corresponding stress fields can be adopted.

Two cases are shown in Fig. 2.79, for which the shear force is transferred in the compression flange (Fig. 2.79a) or in the tension flange (Fig. 2.79b).

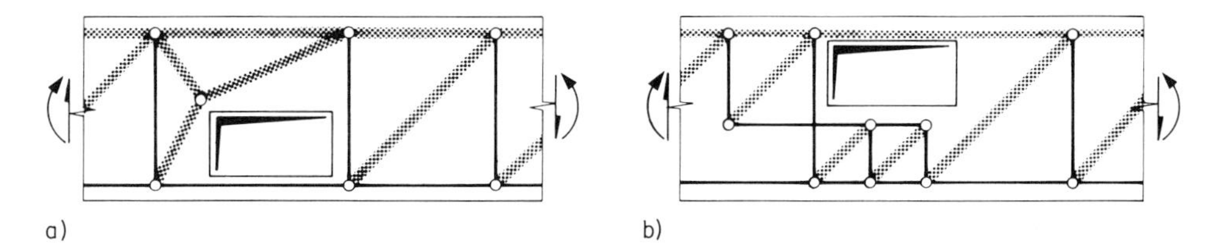

Fig. 2.79: Resultants in the region of holes

With regard to direct or indirect support the same considerations apply as in the case of short columns (section 2.6). If there are several holes separated by a relatively short distance, then connecting beams in the vertical direction result. Fig. 2.80 shows the resultants for such a beam.

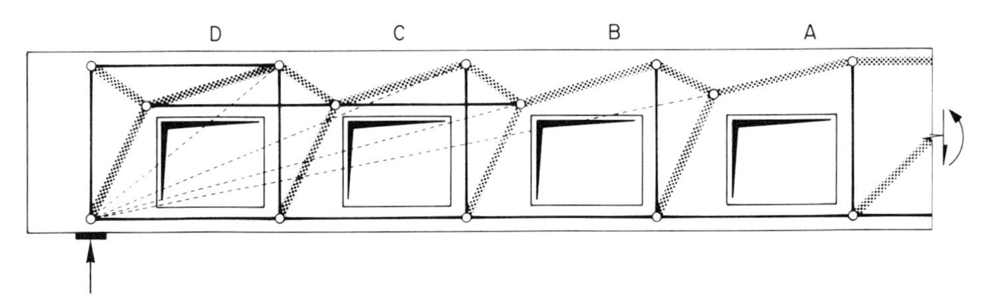

Fig. 2.80: Resultants in beam with multiple holes

For the two horizontal connecting beams *A* and *B* the eccentricity of the resultant compression force is so small that from the statical point of view no reinforcement is necessary in this region. If the resultant lies outside the connecting beams, however, then it has to be forced inside by means of longitudinal reinforcement. For simplification, in Fig. 2.80 at all connecting beams a direct compression strut is assumed. Moreover, it should be observed that the proposed solution for improving the behavior under working loads has to be further investigated along the lines described in section 4.2.

2.9 WALLS

2.9.1 Shear Walls

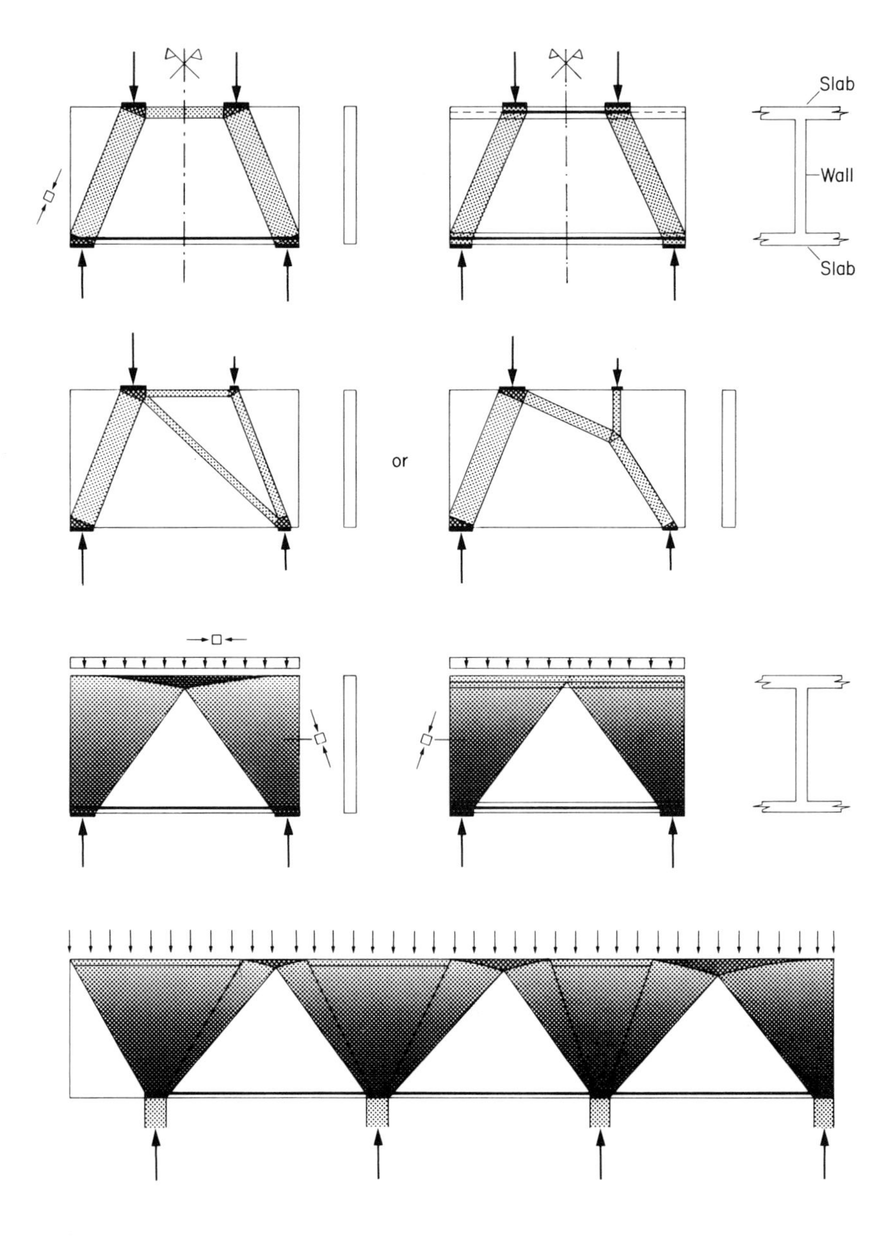

Fig. 2.81: Vertically loaded walls

In this section the investigation is restricted to a few typical plane stress problems. Simple walls loaded vertically were dealt with in section 2.2 as deep beams. Fig. 2.81 shows some examples. For walls fixed monolithically to the floors, the floors can be treated as flange plates.

In the following, typical design cases of shear walls are investigated. In structures, such walls are usually loaded by horizontal forces (wind loading, earthquakes) and vertical loads (dead load, live loads). Fig. 2.82 shows walls which are only loaded in shear. Basically, the vertical reinforcement can be distributed or concentrated. Even in the case of distributed vertical reinforcement, in general, horizontal reinforcement (stringer) is necessary at the upper boundary (Fig. 2.82c).

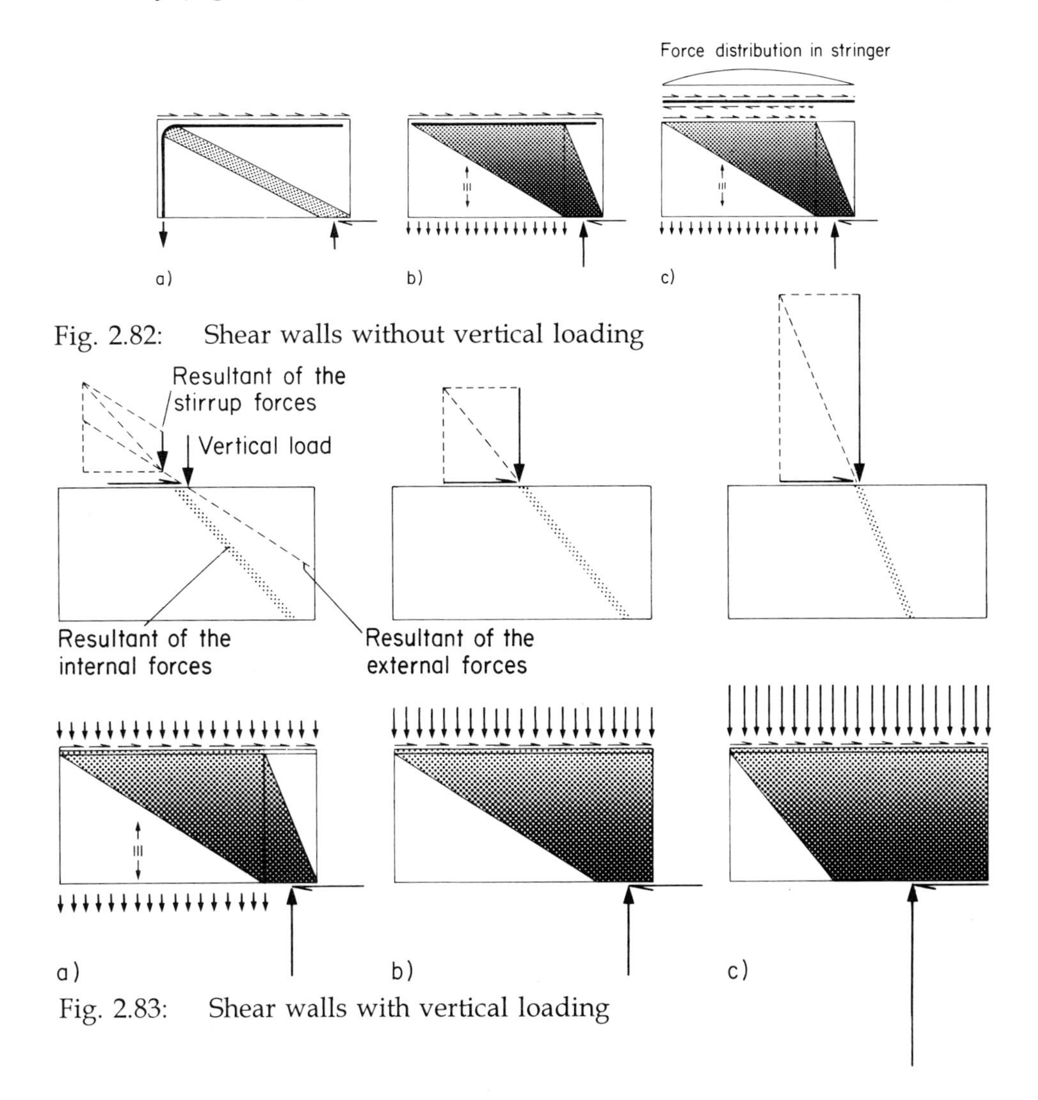

Fig. 2.82: Shear walls without vertical loading

Fig. 2.83: Shear walls with vertical loading

If a vertical load is present it can partially or fully compensate the vertical reinforcement (Fig. 2.83). Fig. 2.83b shows the limiting case, for which no vertical reinforcement is necessary. The position of the resultants of the external forces at the lower edge is determined such that the concrete strength is fully utilized in this region. For the wall shown in Fig. 2.83a the internal resultant is influenced by vertical reinforcement in such a way that this condition is achieved. Similar considerations hold, as shown in Fig. 2.84, for multistory shear walls.

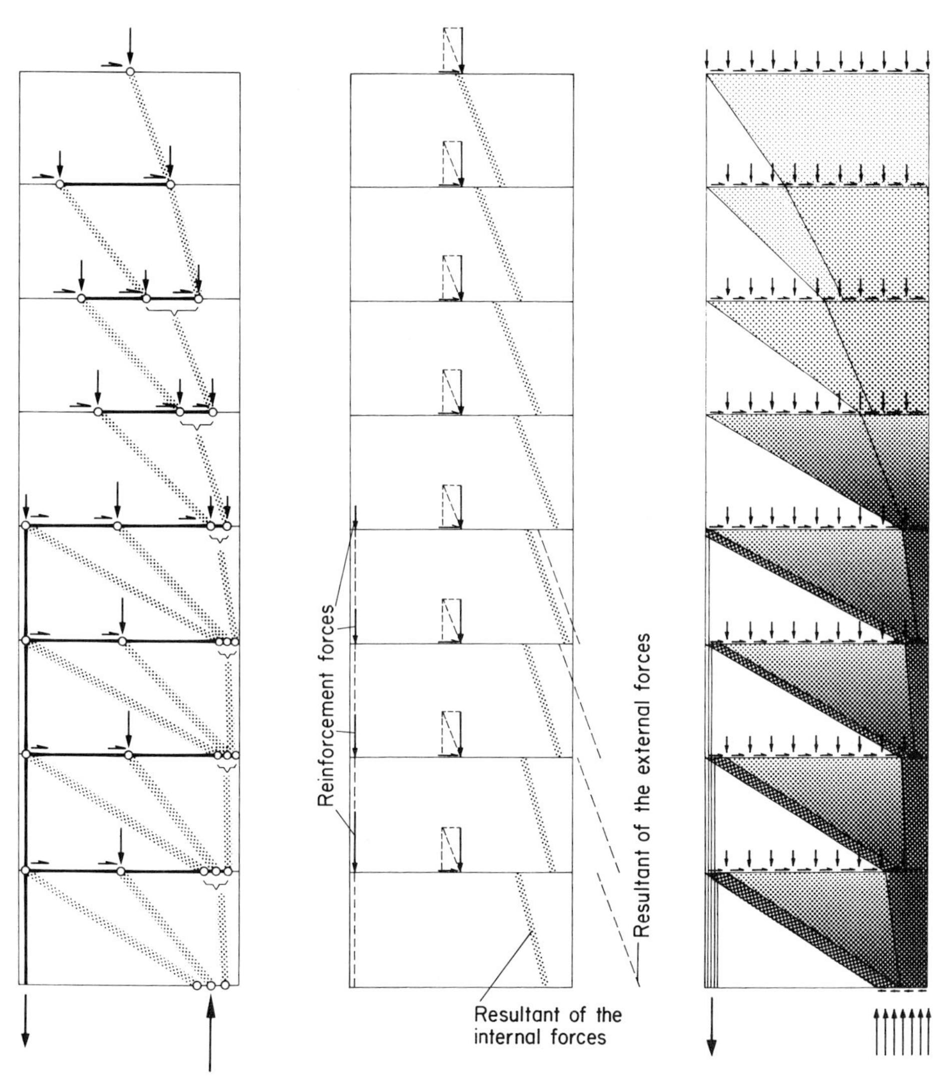

Fig. 2.84: Multistory shear walls, force resultant and stress field

Similar considerations apply to the investigation of walls with irregular geometry or loading. Fig. 2.85 shows such a wall with a hole.

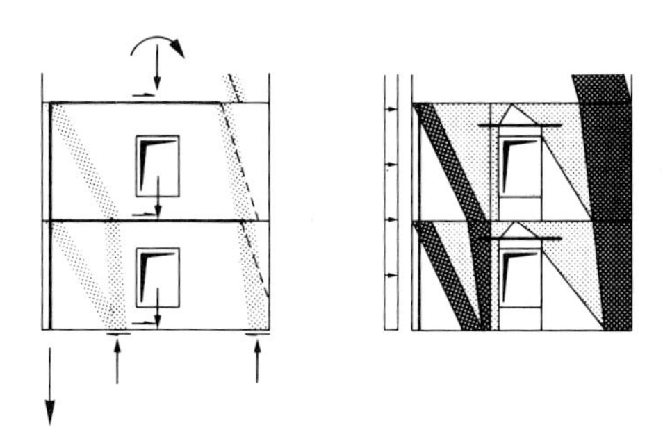

Fig. 2.85: Multistory wall with hole, resultants and stress field

2.9.2 Diaphragms

In members subjected to torsion, the loads and the support forces can be introduced by means of diaphragms (see section 2.4). Fig. 2.86 shows the distribution of the internal forces in diaphragms with the stress fields and the resultants.

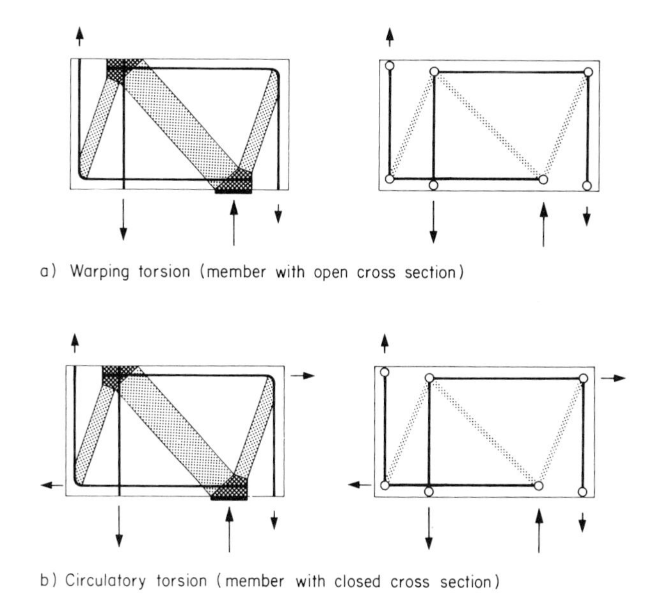

Fig. 2.86: Diaphragm, stress fields and resultants

If the force transfer between the longitudinal wall and the diaphragm is distributed over the boundaries, then the struts shown in Fig. 2.86 can be replaced by compression fields. The development of such a stress field is presented in Fig. 2.87 for the case of circulatory torsion.

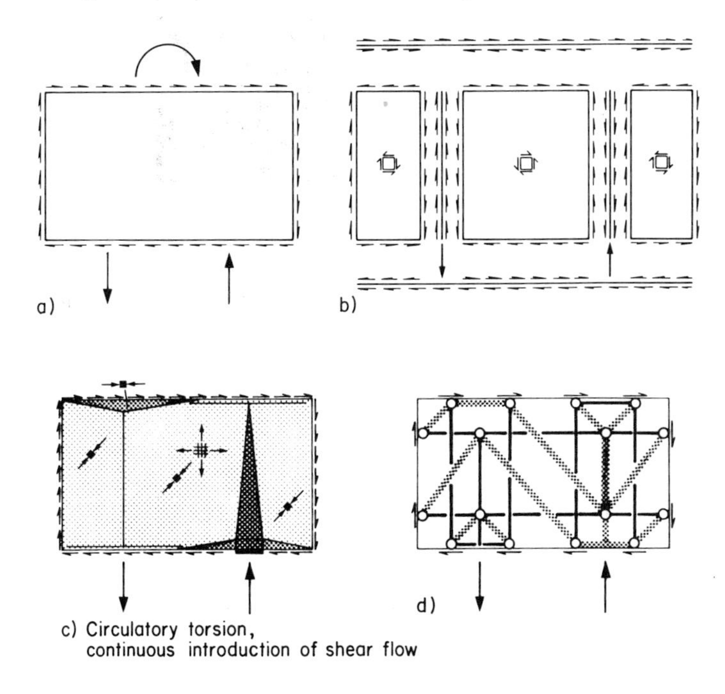

Fig. 2.87: Development of the stress field and resultants

In the case of combined stresses a qualitatively similar field can be developed (Fig. 2.88).

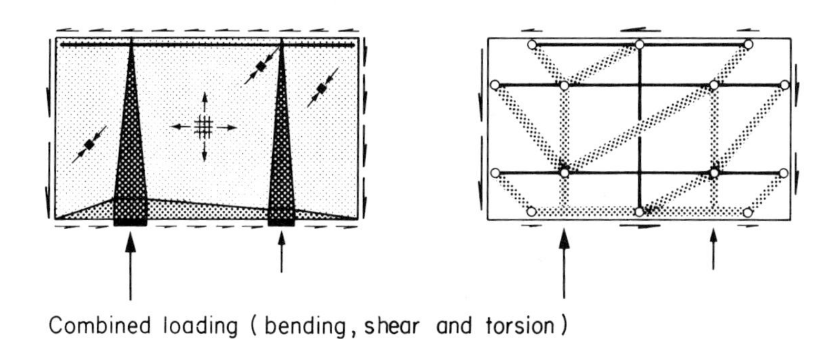

Fig. 2.88: Stress field

If there is a hole in the middle of the described diaphragm, then the distribution of internal forces in the region of the hole can be adjusted locally, as illustrated in Fig. 2.89. The distributed reinforcement forces are acting at the discontinuity lines and have to be anchored behind them.

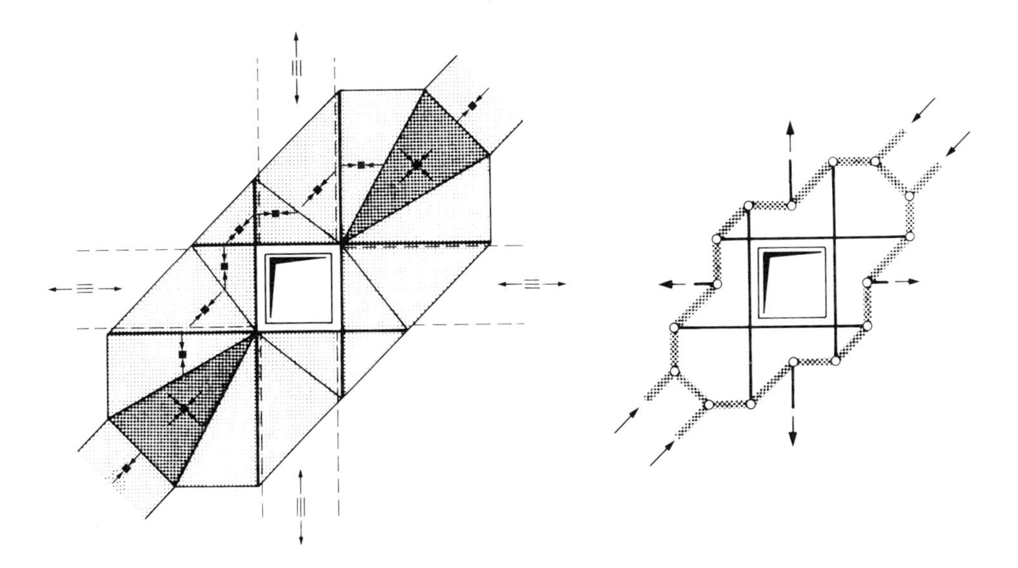

Fig. 2.89: Stress field and internal forces in the region of a hole

2.10 THREE-DIMENSIONAL EXAMPLE

With the aid of a typical example it is shown that the method may also be applied to solid structural members, for which the resultant internal forces do not act in prescribed planes, but rather in an arbitrary manner in space. In Fig. 2.90 a pile cap resting on tree piles, carrying an axial or an eccentric load is shown. The inclinations of the struts are sufficiently large in this case to allow direct support between columns and piles. Thus stirrup reinforcement is unnecessary. The horizontal reinforcement has to be carefully anchored. A possible placement of reinforcement is given in Figs. 2.90c and d.

As in the case of beams, excessively shallow compression diagonals cannot develop due to the possible presence of cracking.

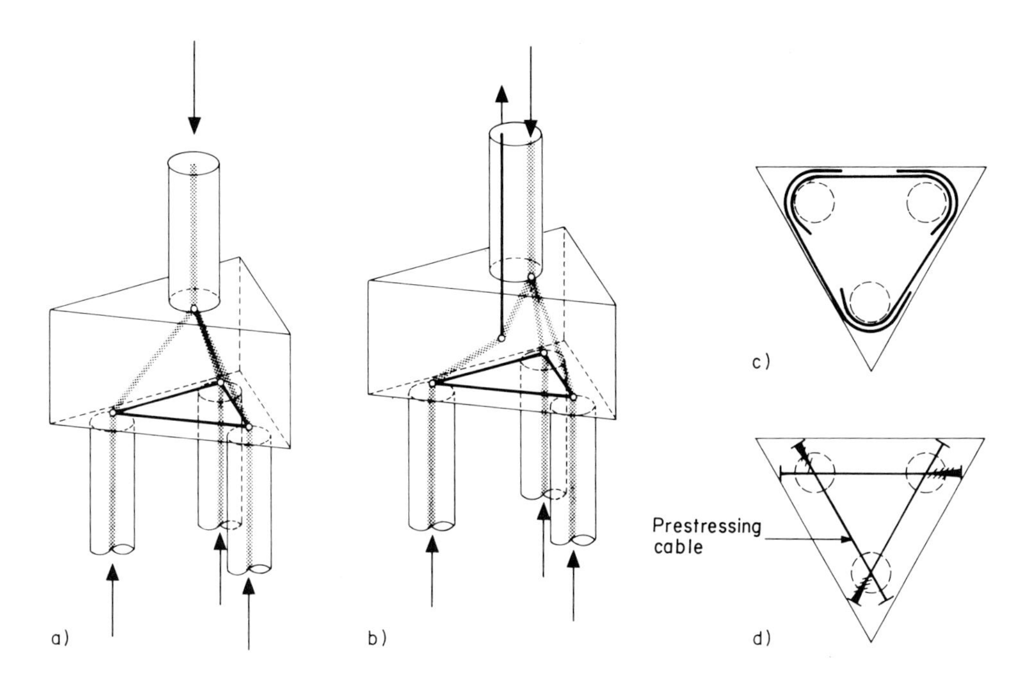

Fig. 2.90: Force resultants in a pile cap, sketch of reinforcement

3 MATERIAL STRENGTHS AND OTHER PROPERTIES

3.1 REINFORCING STEEL

The stress-strain relations for the usual types of reinforcing steels are shown in Fig. 3.1a.

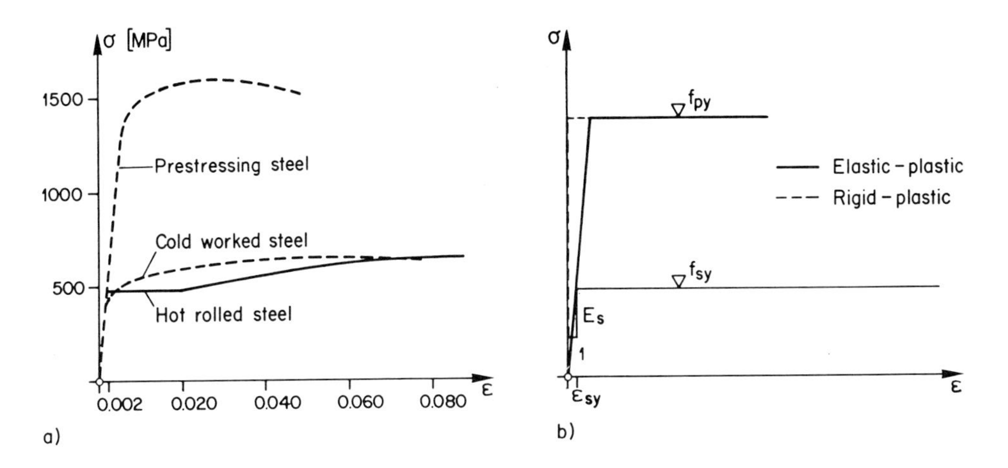

Fig. 3.1: Actual and simplified stress-strain relations

In general, three different regions may be distinguished:

- elastic region (atomic spacing increases in proportion to tensile stress)

- plastic region (deformation of crystal structure with dislocation mechanism)

- hardening region (combination of above two phenomena)

For practical design purposes the simplified stress-strain relations shown in Fig. 3.1b may be used. The very large deformations in the plastic region compared to the yield strain ε_{sy} allow use of the theory of plasticity. If it is assumed that the reinforcing bars take forces in the axial direction only, then a general three-dimensional stress state describing the yield condition in the steel is not required.

3.2 CONCRETE

Since as a rule the concrete is stressed three-dimensionally, it is necessary in this case to introduce a yield condition taking into account the general stress state. As the inelastic strains are not due to dislocation of the crystal structure but to progressive, irreversible damage to the internal material structure, the concrete strength is influenced in addition by different phenomena such as micro-cracking, loading history and interaction with the reinforcement.

Although a material with such properties cannot strictly speaking be classified as ideally plastic, plasticity theory may in general be used, if for the concrete a value of the effective concrete strength f_{ce}, which relates to the above mentioned phenomena, is applied. It should be noticed, that for under-reinforced structural elements (i.e. yielding of the reinforcement precedes crushing of the concrete) the value of f_{ce} has a relatively small influence on the collapse load.

3.2.1 Uniaxial Stress State

Examples of uniaxially stressed concrete elements are the compression zones of beams, plates, and columns as well as the diagonal compression struts of shear walls or elements of smaller aspect ratios. A corresponding state can be produced experimentally in the central zone of uniaxially loaded cylinders or prismatic test specimens. The results of a deformation controlled test for such a case are shown in Fig. 3.2 for three types of concrete.

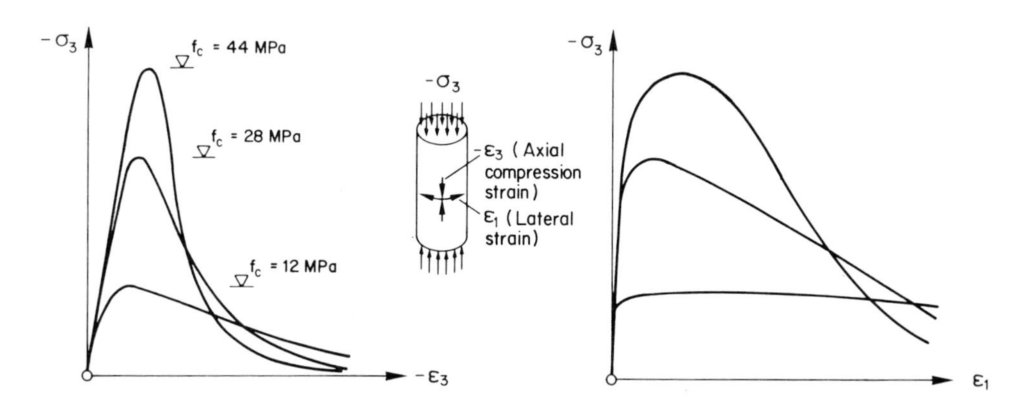

Fig. 3.2: Stress, axial and lateral strain for a uniaxially compressed concrete specimen

Three regions may be observed in the stress-strain diagrams:

- A quasi-elastic region ($0 \leq | \sigma_3 | \leq 0.8 f_c$). The deviation from linear elastic behavior is among other things dependent on the opening of the micro-cracks between the aggregate and the cement matrix. The relationship between lateral and axial strain (Poisson's ratio) is practically constant in this region with a value of 0.15.

- A hardening region with large lateral strains, ($0.8 f_c \leq | \sigma_3 | \leq f_c$). Initiating from the micro-cracks between the aggregate and cement matrix, cracks are formed in the cement matrix parallel to the direction of loading (Fig. 3.3). The axial compression and the lateral strain increase more than linearly.

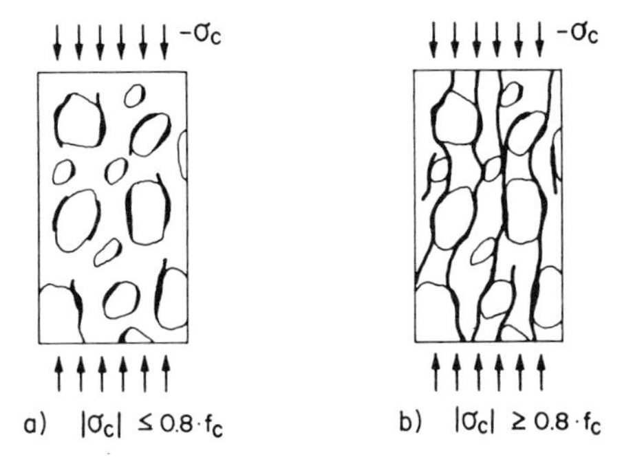

Fig. 3.3: Development of cracks in an uniaxial cylinder test

- A softening region with large lateral strains. The concrete has been broken up by vertical cracks. Some of the resulting pieces are not stable. The stress decreases with increasing deformation.

In order to determine the collapse load of a system with the aid of the theory of plasticity the concrete strength has to be reduced to an effective value f_{ce}. As shown in Fig. 3.4 this reduction depends on the type of system considered. In Fig. 3.4a, for example, a statically determinate system is shown. Since the system's collapse load is given by reaching the cylinder strength, the value $f_{ce} = f_c$ may be adopted.

For statically indeterminate systems (Fig. 3.4b) the strengths of the individual members are not reached simultaneously, but at different deformations of the complete system. For this reason the collapse load of the system can generally not be obtained by summing the collapse loads of the individual members. It can, however, be found by using the theory of plasticity, if the concrete strength is reduced ($f_{ce} \leq f_c$).

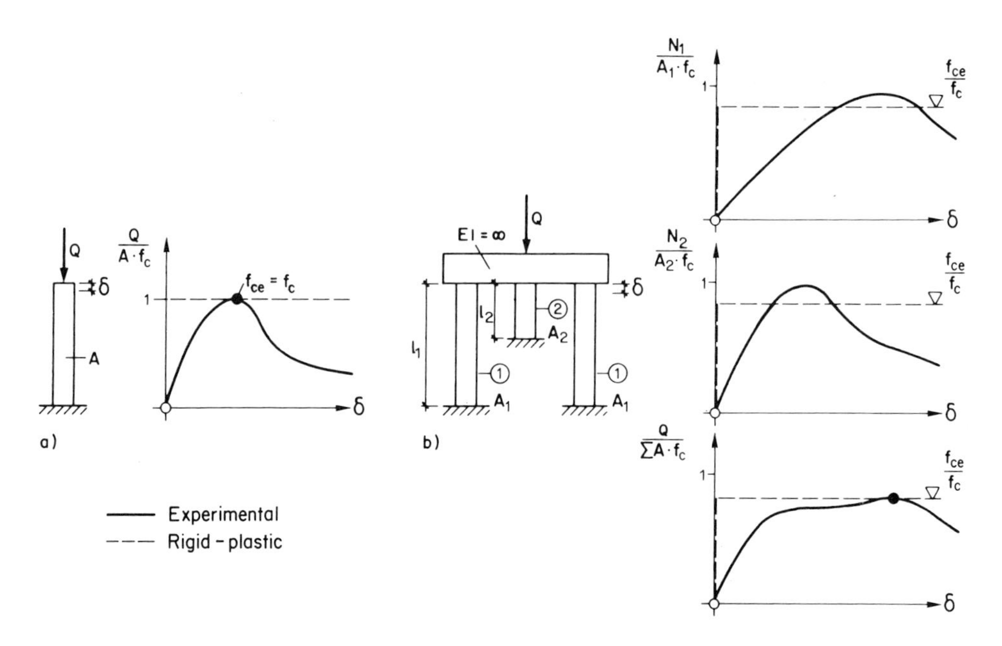

Fig. 3.4: Behavior and strength of statically determinate and indeterminate systems

This reduction depends on the nature of the system (A_1/A_2 , l_1/l_2 for the example given in Fig. 3.4b) and the cylinder strength f_c . The increase of brittleness (significant softening) with increasing cylinder strength as shown in Fig. 3.2 can be taken into account by decreasing to a greater extent the strength of high strength concrete. For typical systems the following values can be adopted:

$$f_{ce} = f_c \qquad \text{for } f_c \leq f_{co} = 20 \text{ MPa} \tag{3.1}$$

$$f_{ce} = f_{co} \cdot (f_c / f_{co})^{2/3} \qquad \text{for } f_c > f_{co} = 20 \text{ MPa} \tag{3.2}$$

A similar relationship has also been derived theoretically by Exner [17]:

$$f_{ce} = 3.2 \cdot f_c^{\,1/2} \qquad (f_{ce}, f_c \text{ in [MPa]}) \tag{3.3}$$

These relationships are valid for systems whose elements exhibit strains of comparable orders of magnitude at collapse. In extreme cases, however, the strength has to be reduced even further or even the strength of the lowly strained elements has to be neglected.

3.2.2 Three-Dimensional Stress State

For an element subjected to additional lateral pressure as shown in Fig. 3.2 the crack development described above is counteracted. The behavior of a concrete element loaded in such a way is illustrated in Fig. 3.5. The lateral pressure can be applied either as an external load or by inhibiting lateral strain (e.g. by means of spiral reinforcement).

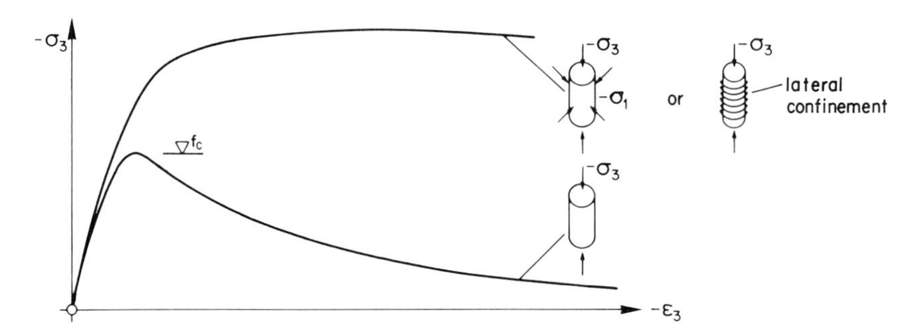

Fig. 3.5: Stress-strain relation for three-dimensionally and uniaxially compressed concrete

The increase of strength can be described according to the Mohr-Coulomb hypothesis (Fig. 3.6).

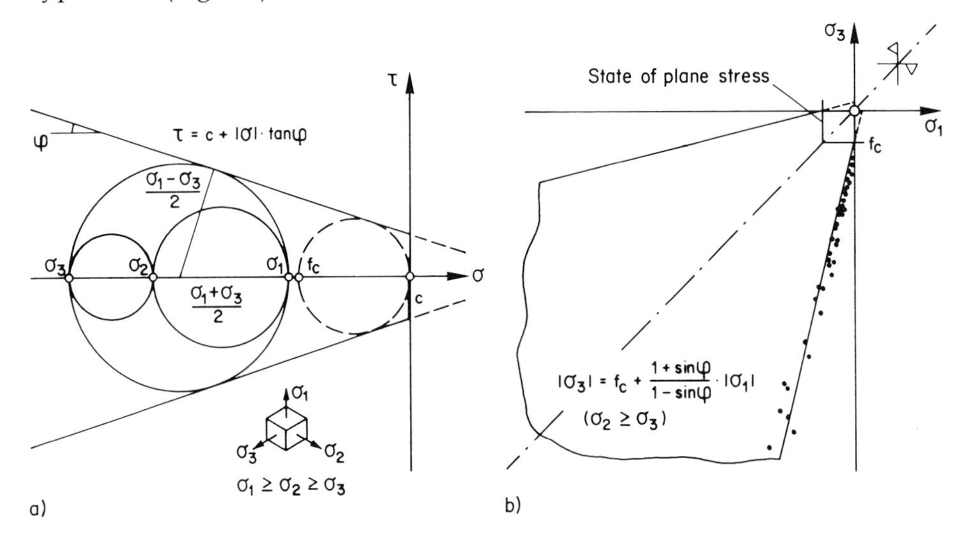

Fig. 3.6: Yield condition according to the Mohr-Coulomb hypothesis
a) Mohr's representation, b) Principal stress space representation with test results [18]

As shown, by a comparison with the test results included in Fig. 3.6b, an appropriate value of the parameter $\varphi \approx 37°$ may be assumed. Thus the increase of strength due to a hydrostatic stress state is $\Delta f_{ce} = 4 \cdot | \sigma_1 |$.

It is evident from Fig. 3.5 that the lateral confining pressure not only leads to an increase in strength, but also to an increase in ductility. This phenomenon may be explained in terms of different fracture processes. In the case of large lateral confining pressures failure is initiated primarily by slip along an inclined failure surface (Fig. 3.7) and not as in the case of uniaxial compression by instability due to longitudinal splitting.

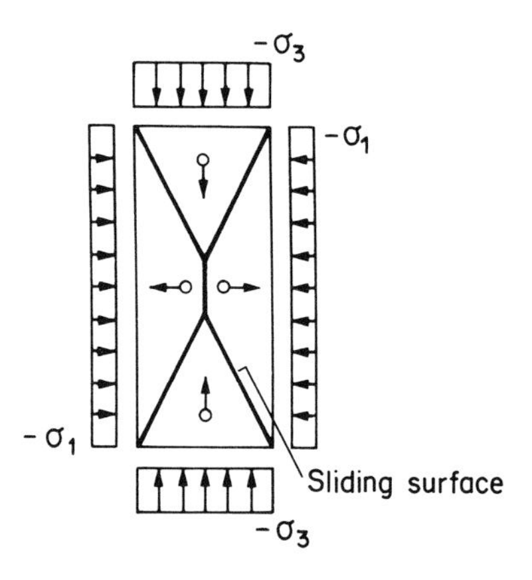

Fig. 3.7: Failure mechanism in case of lateral confinement

The increase in ductility can be taken into account by simply assuming that the increase Δf_{ce} corresponds to an ideal ductile behavior. Thus for f_{ce} the following relationships apply:

$$f_{ce} = f_c + 4 \cdot | \sigma_1 | \qquad \text{for } f_c \le f_{co} = 20 \text{ MPa} \qquad (3.4)$$

$$f_{ce} = f_{co} \cdot (f_c / f_{co})^{2/3} + 4 \cdot | \sigma_1 | \qquad \text{for } f_c > f_{co} = 20 \text{ MPa} \qquad (3.5)$$

There also exists in the concrete cube test a relatively complicated three-dimensional stress state caused by the prevention of lateral strain at the loading planes (Fig. 3.8). For this reason the strength obtained in the cube test is greater than f_c ($f_{cw} \approx 1.25\, f_c$) and is therefore not representative of the uniaxial stress state.

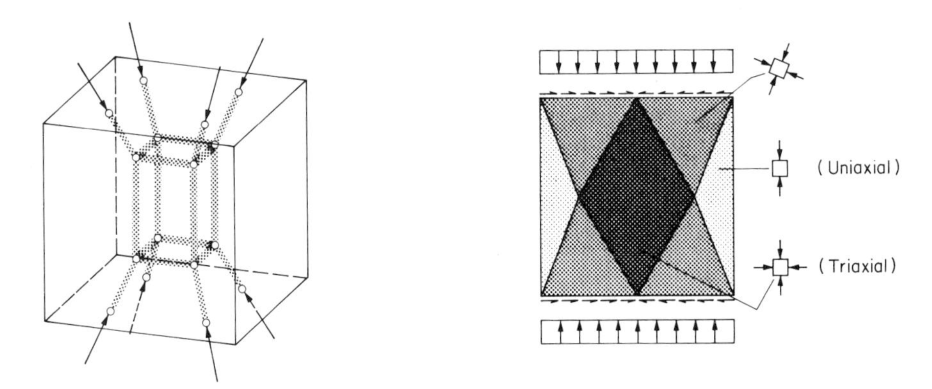

Fig. 3.8: Schematic representation of the internal forces and stress field

A similar behavior may be found in locally highly stressed zones, whose size is about the same as that of the lateral dimension. Here it is a case of restricted lateral strains due to neighboring unstressed zones of concrete (Fig. 3.9). Should there be in addition some stirrup reinforcement then the concrete strength may be considerably greater than f_c.

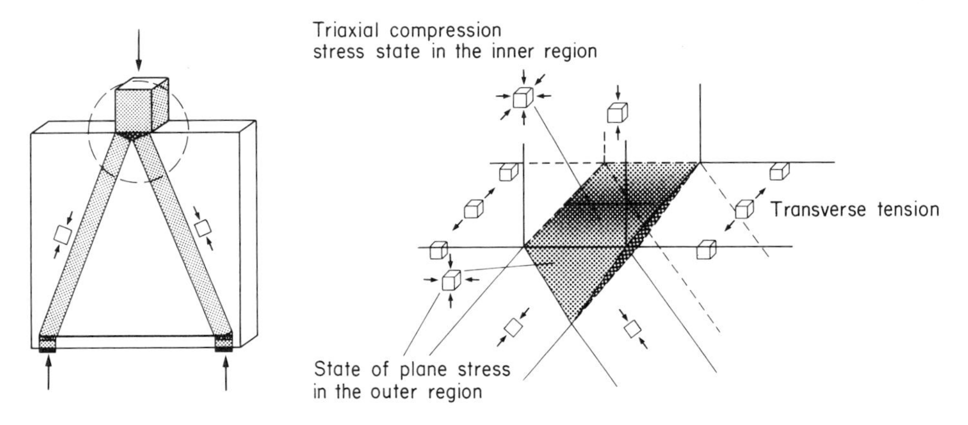

Fig. 3.9: Lateral strains impeded in locally loaded zones

3.2.3 Concrete with Imposed Cracks

For the stress fields developed in chapter 2 uniaxial stress fields in concrete were often superimposed by a strain state imposed by the elongation of the tensile reinforcement. Such a case with the reinforcement running normal to the compression field is illustrated in Fig. 3.10.

The deformation of the reinforcement produces cracking in the concrete. Thus there results a similar concrete strain softening as in the case of the lateral deformation of a uniaxially loaded concrete element.

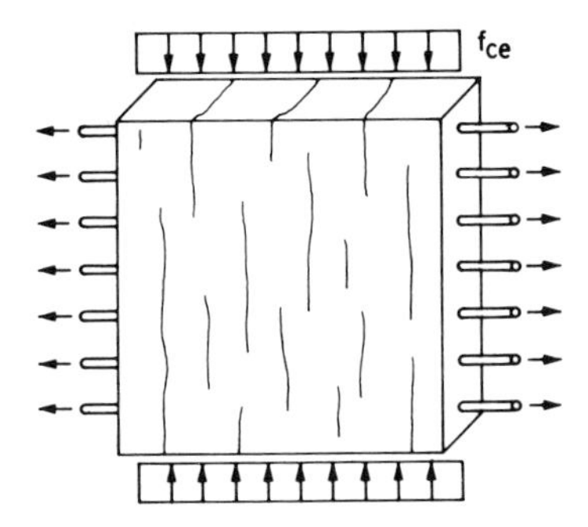

Fig. 3.10: Concrete compression field with transverse reinforcement under tension

The stress in the concrete as a function of lateral strain with and without transverse reinforcement is shown in Fig. 3.11. Since the deformations are not quantified in the plastic design method (rigid-plastic assumption), this phenomenon can be accounted for only indirectly by taking an effective strength f_{ce} which is less than f_c. Hereby, it is accounted for that the behavior of concrete is not ideally plastic.

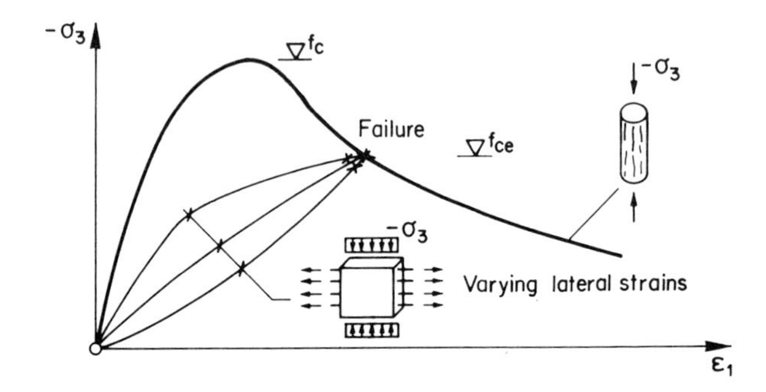

Fig. 3.11: Influence of laterally imposed strains on stress-strain curve of concrete

As is evident from a comparison of the stress-strain curves for different values of cylinder strength f_c (Fig. 3.2), for a given lateral strain ε_1 the ratio f_{ce}/f_c is also dependent on the cylinder strength f_c.

For moderate lateral strains ($\varepsilon_1 < 3\%$) the following values can be adopted for the case of transverse reinforcement normal to the concrete compression field:

$$f_{ce} = 0.8 \cdot f_c \qquad \text{for } f_c \leq f_{co} = 20 \text{ MPa} \tag{3.6}$$

$$f_{ce} = 0.8 \cdot f_{co} \cdot (f_c / f_{co})^{2/3} \qquad \text{for } f_c > f_{co} = 20 \text{ MPa} \tag{3.7}$$

If the transverse reinforcement is not normal to the concrete compression field then there is additional strain softening in the concrete when a crack opens (Fig. 3.12).

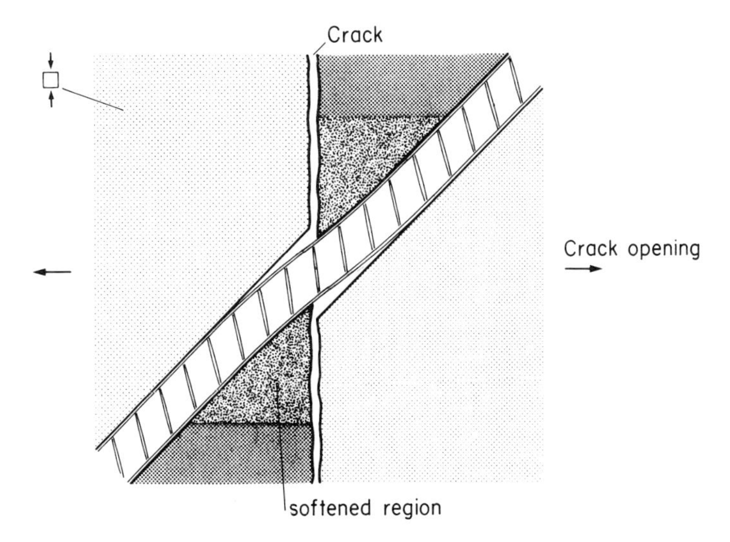

Fig. 3.12: Interaction between concrete and reinforcement during crack opening

A transverse tensile stress results from transverse bending of the reinforcing bars, which leads to spalling of the concrete cover near failure. Normally the following values apply:

$$f_{ce} = 0.6 \cdot f_c \qquad \text{for } f_c \leq f_{co} = 20 \text{ MPa} \tag{3.8}$$

$$f_{ce} = 0.6 \cdot f_{co} \cdot (f_c / f_{co})^{2/3} \qquad \text{for } f_c > f_{co} = 20 \text{ MPa} \tag{3.9}$$

A comparison with the test results is represented in [19].

3.2.4 Cracks: Aggregate Interlock

Both the load history and the redistribution of internal forces lead in some cases to a force transfer at an open crack. If the concrete compression field runs oblique to the 'plane of the crack', then this force transfer results from the so-called aggregate interlock action.

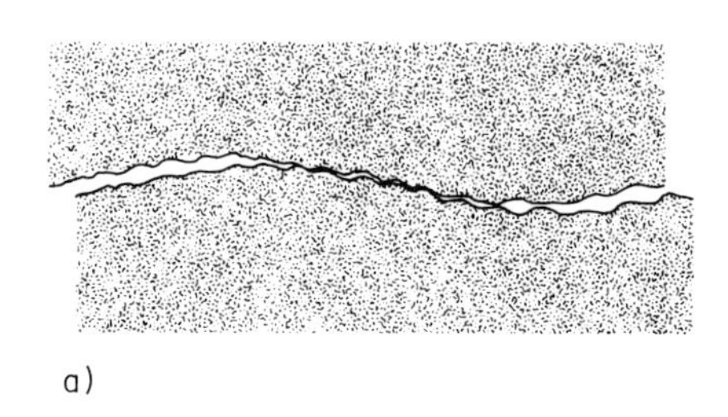

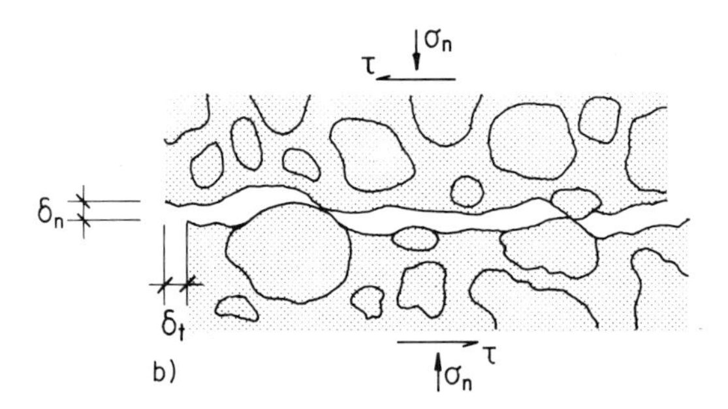

Fig. 3.13: a) Macro-interlocking (crack shape)
b) Micro-interlocking (aggregate interlock)

For small crack openings δ_n and sufficiently large displacements along the crack δ_t the micro-interlocking acts on the total crack area (Fig. 3.13b), whereas for large crack openings the force can only be transmitted via macro-interlocking in local zones (Fig. 3.13a).

With regard to practical design it should be mentioned that in general the internal force transfer at the failure state can only be guaranteed if the crack widths are kept small by means of reinforcement.

The effective strength f_{ce} for the cases discussed above are summarized in Fig. 3.14.

TYPE OF LOADING		EFFECTIVE STRENGTH f_{ce} $f_{ce} \leq f_{c0} = 20 MPa$	$f_{ce} \geq f_{c0} = 20 MPa$	APPLICATIONS
Uncracked, with lateral compression or lateral confinement	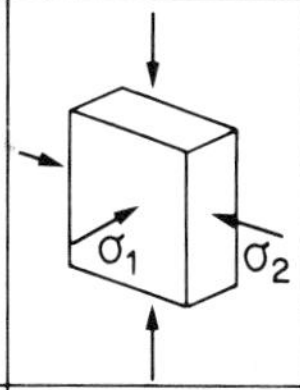 σ_1 σ_2	$f_c + 4\,\lvert\sigma_1\rvert$	$f_{c0}(\frac{f_c}{f_{c0}})^{\frac{2}{3}} + 4\,\lvert\sigma_1\rvert$	– Triaxial compression – Lateral confinement (introduction of concentrated loads, column with spiral reinforcement....)
Uncracked, uniaxial compression	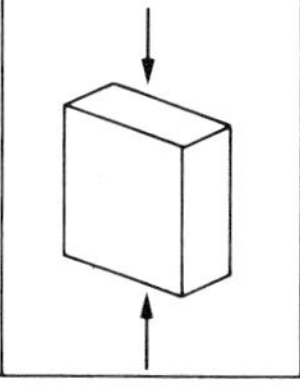	f_c	$f_{c0}(\frac{f_c}{f_{c0}})^{\frac{2}{3}}$	– Pure compression (columns....) – Bending, bending + axial force (beams and slabs) – Walls
Uniaxial compression, axial cracks, laterally imposed strains	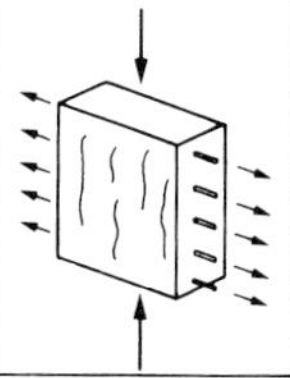	$0.8\,f_c$	$0.8\,f_{c0}(\frac{f_c}{f_{c0}})^{\frac{2}{3}}$	– Walls – Beams with imposed lateral strain – Slabs
Uniaxial compression, axial cracks, diagonally imposed strains	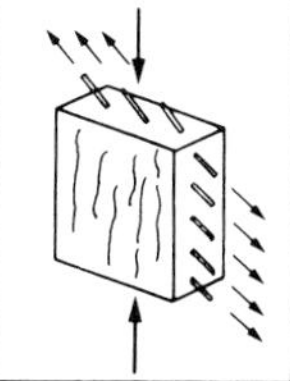	$0.6\,f_c$	$0.6\,f_{c0}(\frac{f_c}{f_{c0}})^{\frac{2}{3}}$	– Beams subjected to shear, torsion – Walls – Slabs with large twisting moments
No prevention of large cracks	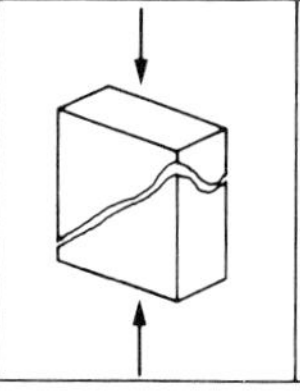	Theory of plasticity not applicable, see Chapter 4		– Beams and walls with insufficient minimal reinforcement – Slabs subjected to large shear forces (punching)

Fig. 3.14: Effective concrete strength

3.3 FORCE TRANSFER REINFORCEMENT - CONCRETE

Since the breakdown of such force transfer is normally characterized by brittle behavior this type of failure should not be governing in design. It is not the aim here to give quantitative information for design purposes, but rather to describe qualitatively the transfer mechanism and the important factors influencing strength.

3.3.1 Anchorage of Reinforcing Bars

The transfer of the shear forces between the reinforcement and the concrete for smaller loads is due primarily to bond action (molecular bonding and micro-interlocking). After the bond strength has been exceeded the force transfer is due to the ribs in the reinforcing steel. For a description of the distribution of the internal forces at the failure state the stress field for the force transmission in the compression flange of T-beams can be adopted and extended. The axisymmetric stress fields obtained in this way are a simplification of the actual stress fields. In Fig. 3.15a it may be seen that the deviation of the diagonal compression field is achieved by means of a ring-shaped concrete tensile field, while in Fig. 3.15b it is by means of circumferential or spiral reinforcement.

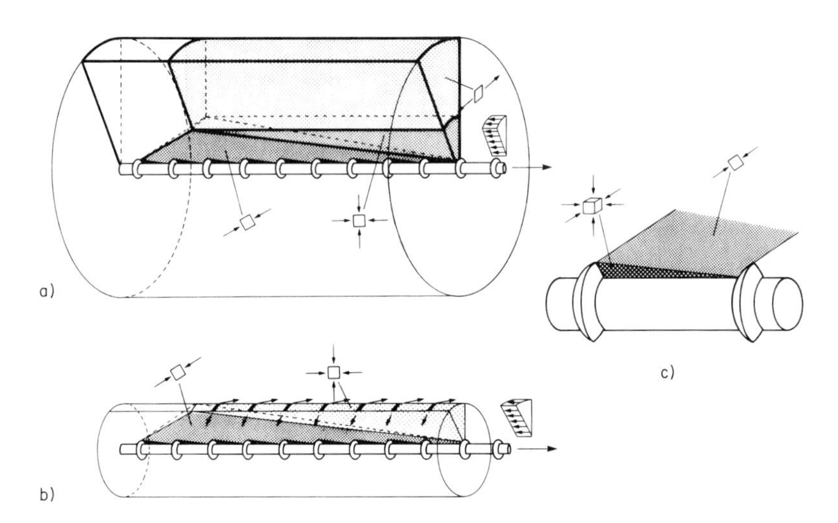

Fig. 3.15: Stress fields describing the stress state in the anchorage zone
a) Taking up the spreading force with a concrete tension ring,
b) Taking up the spreading force with a confining reinforcement,
c) Detail

A premature anchorage failure occurs when either the concrete tension ring fails or the spiral reinforcement yields. The compressed concrete in the vicinity of the reinforcing bars (diagonal concrete compression field) is, as a rule, not critical, because it is stressed three-dimensionally.

If there is no spiral reinforcement then the failure load depends on the anchorage length as well as the concrete tensile strength and the thickness of the concrete tensile ring. A theoretical evaluation of the stress field according to Fig. 3.15a results in a linear relationship between the bond strength and the thickness of the concrete tension ring (concrete cover). The test results included in Fig. 3.16 taken from [20, 21, 22] confirm this relationship. It should be noted that a possible crack in the area of the tension ring can have a negative influence on the bond strength.

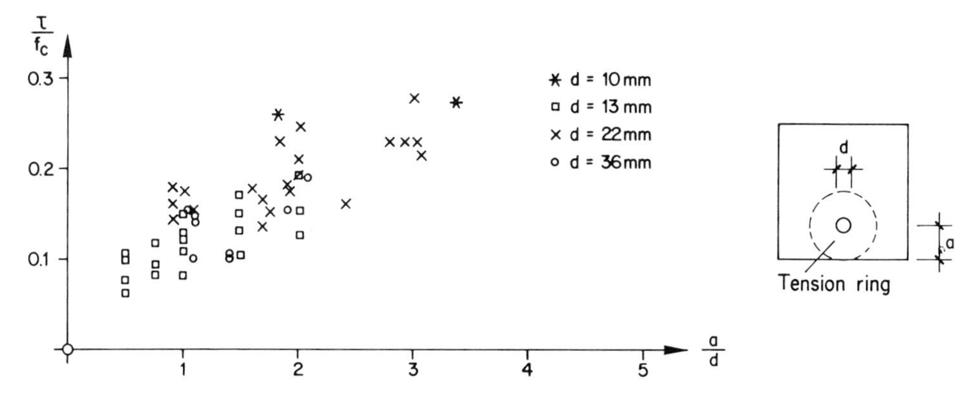

Fig. 3.16: Bond strength as a function of the concrete cover tests from [20,21,22] ($\tau = N_s / \pi \cdot d \cdot l$, N_s = force in reinforcing bar, l = length of anchorage zone)

In Fig. 3.16 the bond strength has been normalized by means of the value $\tau_0 = f_c$. The scatter may be reduced by defining τ_0 as a nonlinear function of f_c. The graphs using τ / f_c shown in this section are therefore not applicable to high strength concrete.

Circumferential reinforcement or confining pressure have the same effect on the bond strength. In order to make the spiral reinforcement fully effective, in the anchorage zone it must surround the whole bar with an adequate spacing. Normally, however, transverse reinforcement is arranged near the surface of the concrete which is in contact with the bar that has to be anchored, so that it is not possible to have an axisymmetric stress field. In this case the stress field shown in Fig. 2.38 can be adopted which was previously used to describe the force distribution in the compression flange of the T-beam (Fig. 3.17).

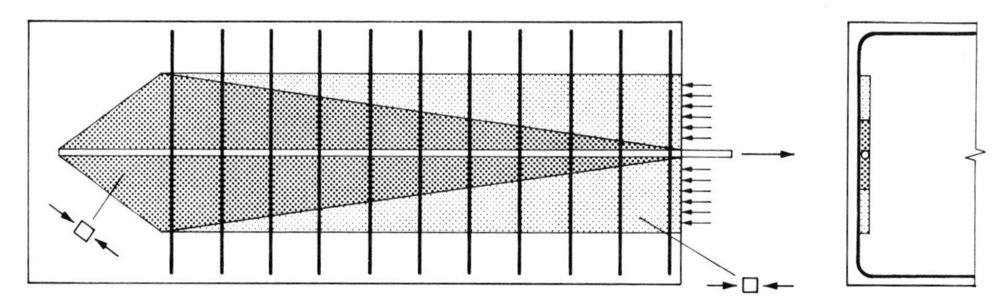

Fig. 3.17: Anchorage zone: plane stress field

In Fig. 3.18 the experimentally determined dependence of the bond strength on the amount of transverse reinforcement are shown (test results taken from [23]). For large amounts of transverse reinforcement the concrete in an inclined compression field is the critical factor.

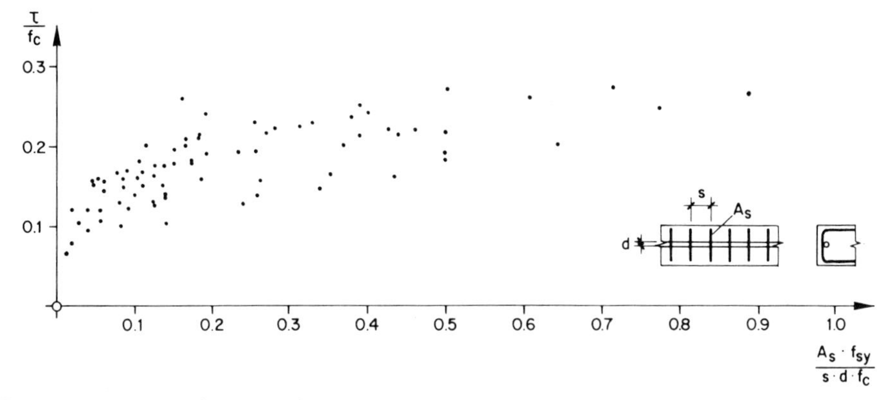

Fig. 3.18: Bond strength as a function of the amount of transverse reinforcement

3.3.2 Splices of Reinforcement

The description of the stress state in the overlapping region of spliced reinforcing bars can be simplified using the stress fields shown in Fig. 3.19. Between the two reinforcing bars a diagonal stress field can develop, whose lateral components can be taken up with the aid of a ring-shaped concrete tension field (Fig. 3.19a) or by means of spiral reinforcement. In either case a three-dimensional stress field results. If the transverse reinforcement, however, is arranged near the concrete surface and is in contact with the spliced bars, then the internal forces correspond to the plane stress field shown in Fig. 3.19b.

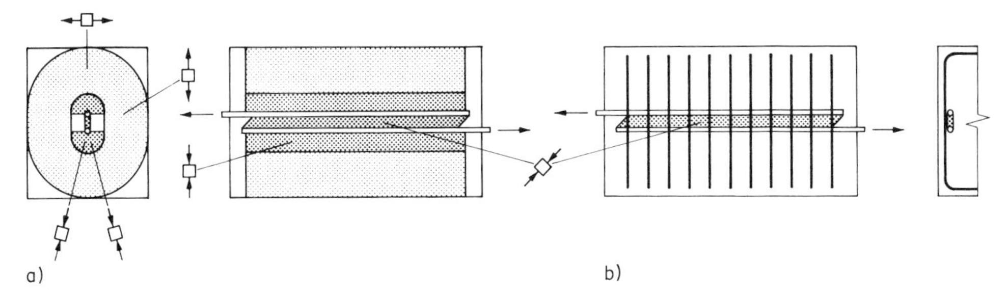

Fig. 3.19: Stress field in splices: a) field with concrete tension ring, b) plane stress field

As is evident from Fig. 3.20a, as in the case of anchoring reinforcing bars, for overlapping bars without transverse reinforcement the failure load depends on the length of overlapping and above all on the thickness of the concrete tension ring (i.e. on the concrete cover) and the concrete strength. Also, in the case of plane transverse reinforcement the dependence of the bond strength on the amount of transverse reinforcement (Fig. 3.20b) is similar to the case of anchoring reinforcing bars.

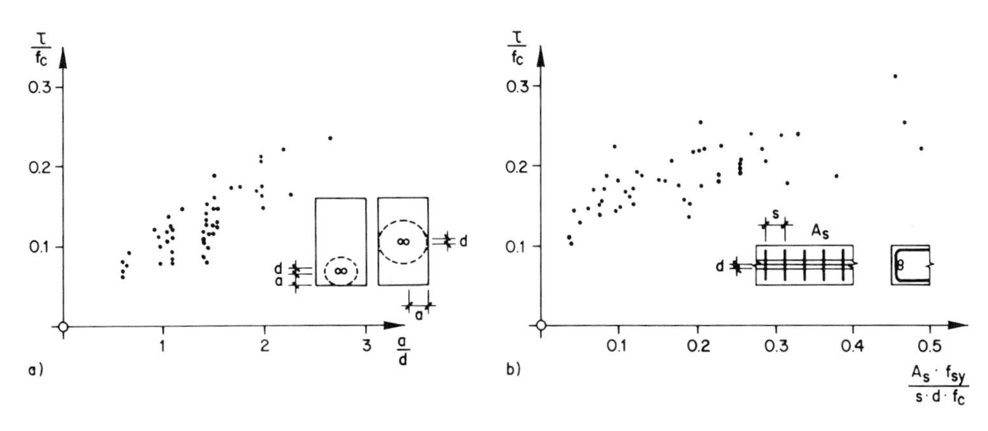

Fig. 3.20: Experimentally determined bond strength in splices as a function of a) concrete cover, b) amount of transverse reinforcement (test results taken from [24, 25, 26, 27, 28, 29, 30])

3.3.3 Force Deviations

For the stress fields dealt with in chapter 2, joint details are used in which the forces in the struts are in equilibrium with the forces in the steel reinforcement.

Fig. 3.21 shows two typical nodal regions whose structural behavior and strength are treated in this section.

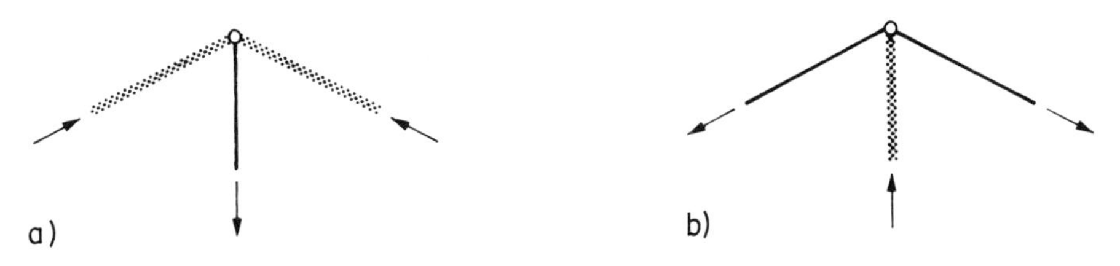

Fig. 3.21: Resultants in nodal region

For the node shown in Fig. 3.21a a strut force is deviated by means of reinforcement. Depending on the space available, the reinforcement force is built up inside or outside the nodal region (Fig. 3.22).

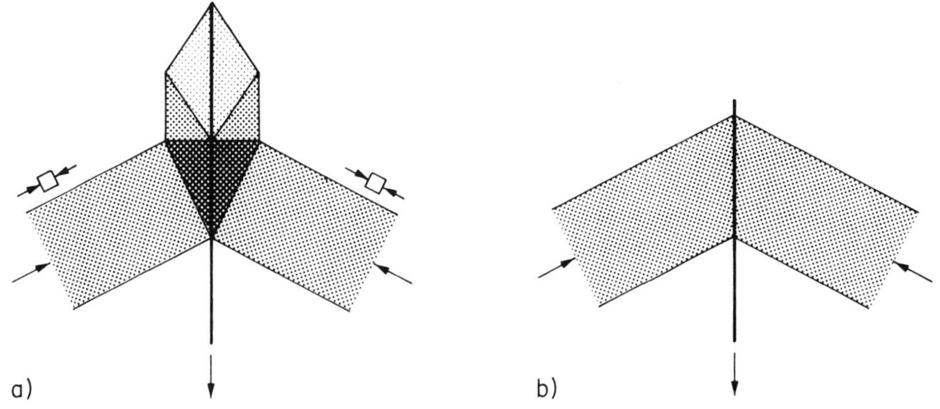

Fig. 3.22: Force transfer reinforcement - concrete outside (a) and inside (b) the nodal region

These two alternative methods for deviating strut forces were introduced in chapter 2 (Fig. 2.8). In the first case, in which the force transfer between the concrete and the reinforcement is effected outside of the nodal region, the structural behavior and the strength can be treated analogously to the case of anchored reinforcing bars.

If, on the other hand, the force transfer is effected inside the nodal region, then the structural behavior depends above all on the geometry of the node itself as well as the arrangement of the reinforcement. Fig. 3.23 shows the resultants of the internal forces for a single node with a straight reinforcing bar as well as the possible modes of failure.

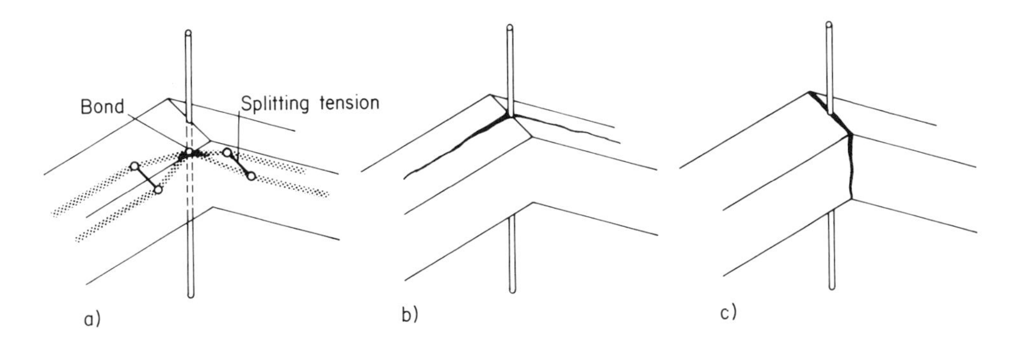

Fig. 3.23: Resultants of the internal forces and possible failure modes

The mode of failure shown in Fig. 3.23b corresponds to a failure of the zone subjected to transverse tension. The strength in this case is strongly dependent on transverse reinforcement, if present, or on a lateral confining pressure.

The other mode of failure (Fig. 3.23c), on the other hand, may be traced back to a failure of the bond between the concrete and the reinforcing steel. Here too the structural behavior is related to that discussed in section 3.3.1.

The component of the strut forces acting normal to the reinforcement produces a lateral pressure, which is favorable to the structural behavior. This relationship is clearly seen from an examination of the test results shown in Fig. 3.24 (after [31]) and was verified by Hess [32] with the aid of the kinematic method of the theory of plasticity.

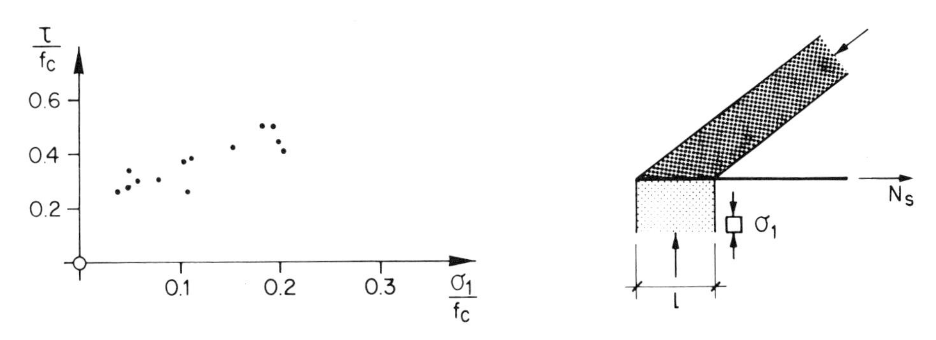

Fig. 3.24: Dependence of the bond strength on the lateral confining pressure [31], ($\tau = N_s / \pi \cdot d \cdot l$, d = diameter of bar)

Other ways of arranging the reinforcement are shown in Fig. 3.25 together with the distribution of the internal forces.

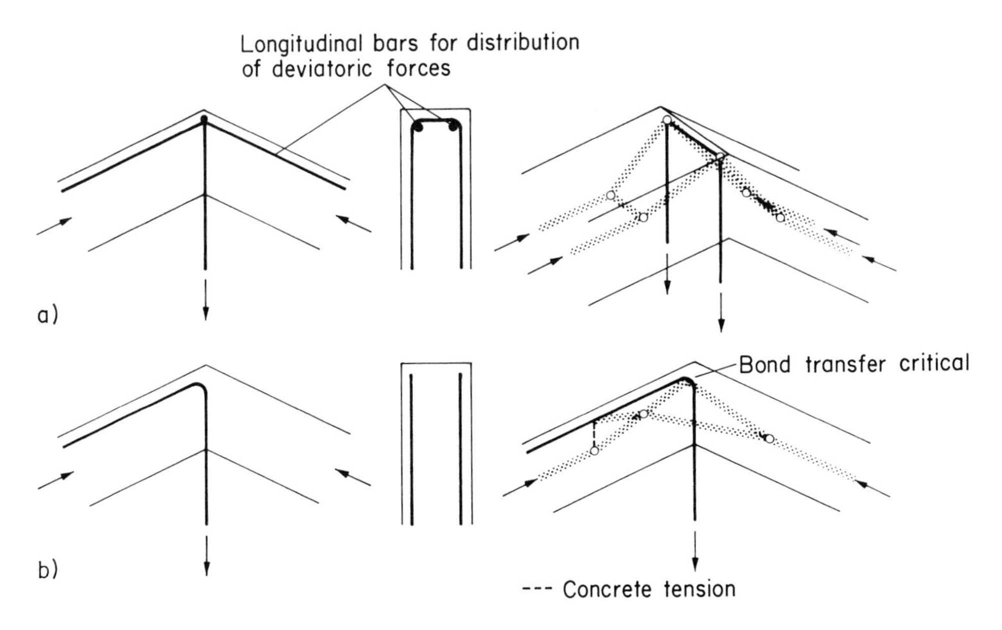

Fig. 3.25: Different arrangements of reinforcement and the associated internal forces

In the case of loops (Fig. 3.25a) the force transfer between concrete and steel reinforcement does not result primarily through bond, but by means of force deviation in the bent. Since these forces act in a relatively concentrated manner, the mode of failure shown in Fig. 3.23c can develop. Large diameter longitudinal bars distribute the deviated forces and hinder a separation of the concrete in the nodal region.

When anchoring with a bent steel bar the structural behavior of the connecting strut is improved, but that of the other becomes unfavorable compared with the solution with straight bars (Fig. 3.22b). Fig. 3.25b shows that the force transfer between the concrete and the reinforcing steel occurs essentially in the bent, on the left side without any problem by means of deviatoric forces and on the right side by bond with an unfavorable slanting angle between the curved reinforcing bar and the strut.

The structural behavior for the solutions shown in Fig. 3.25 is strongly dependent on the arrangement of the reinforcement.

In Fig. 3.26 analogous arrangements of reinforcement with large radii of curvature of the bars in comparison with the dimension of the strut are shown.

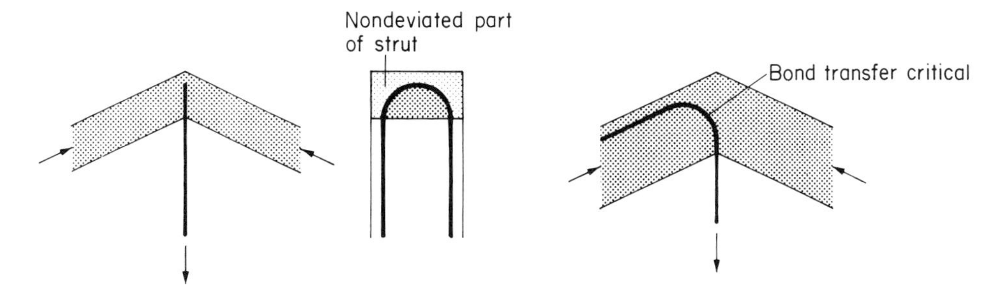

Fig. 3.26: Problematic deviation for large radii of curvature of the bars

For joints of frames with compression outside and tension inside the size of the strut is usually so small that with the above arrangement of reinforcement a complete concentrated force deviation is not possible (see section 2.7.2) and can only be achieved with the help of externally applied anchor plates.

These considerations are also applicable for nodes with two tensile forces and one compressive force according to Fig. 3.21b. For symmetric nodes the force transfer is accomplished only by force deviation. Fig. 3.27 shows a plane stress field based on chapter 2 and a three-dimensional sketch of the distribution of the internal forces.

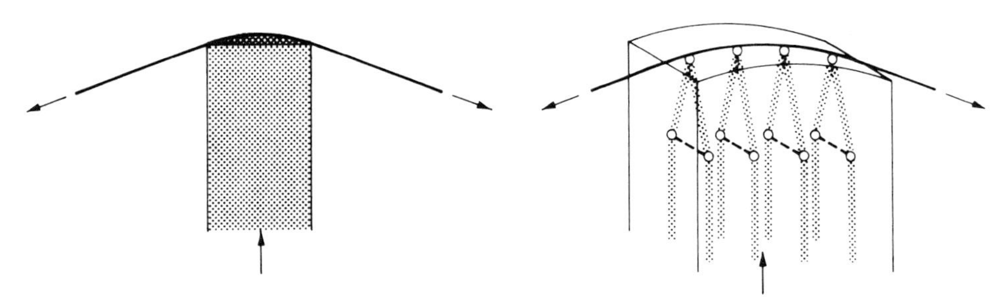

Fig. 3.27: Plane stress field and resultants

It is evident that the region immediately below the curved reinforcing bar is subjected to three-dimensional compression. The local strength is thus higher than the cylinder strength and normally not governing. The cause of collapse is tensile splitting in the transverse direction.

The strength of the nodal connection is above all dependent on the concrete cover of the bar in the tension strut in the transverse direction, on the transverse reinforcement, if present, and on the lateral confining pressure.

In Fig. 3.28 the dependence of the local concrete strength on the concrete cover of the curved reinforcing bar is given (taken from [33]). From this figure it may be seen that either the radius of the reinforcing bar on the concrete cover may govern the design.

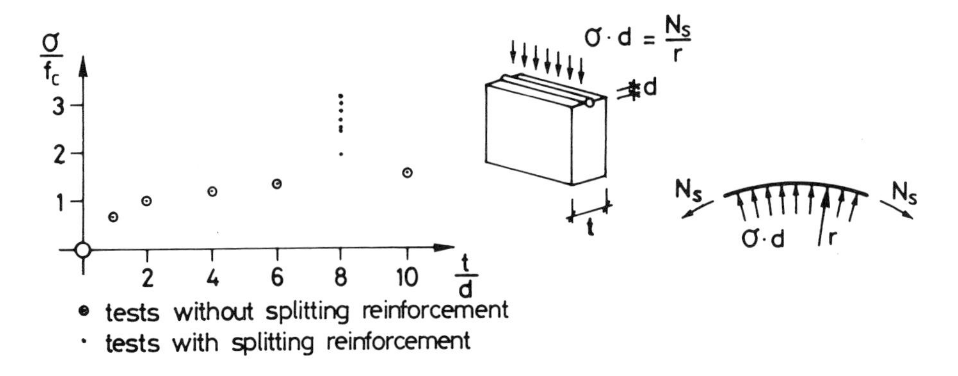

Fig. 3.28: Influence of the concrete cover

For non-symmetrical nodes the force in the strut can be resolved into two components, in such a way that the one component is in equilibrium with the deviatoric forces and the other with the bond stresses of the reinforcing bar (Fig. 3.29a). If the bond strength in the nodal region is not sufficient then this force transfer has to take place in the neighboring region (Fig. 3.29b).

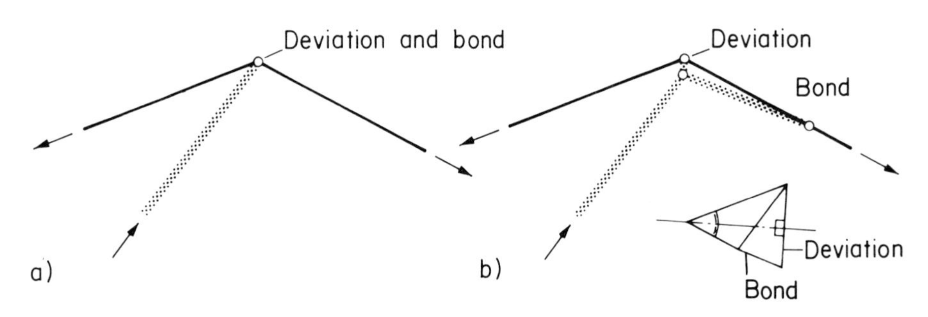

Fig. 3.29: Sketch of force transfer

4 ADDITIONAL CONSIDERATIONS FOR THE DEVELOPMENT OF STRESS FIELDS

4.1 REMARKS ON PLASTIC DESIGN APPLIED TO REINFORCED CONCRETE

Since in statically indeterminate structures more than one distribution of internal forces, which fulfils equilibrium is possible, further criteria are possibly necessary in design to guarantee a satisfactory structural behavior at working load. To explain this observation, the behavior of a statically indeterminate structure is considered. Elastic and plastic design methods are compared. Then the behavior at working load of an elastically designed structure is compared to that of a plastically designed structure. From these investigations criteria are obtained for the choice of suitable stress fields.

4.1.1 Behavior of Statically Indeterminate Beams

Fig. 4.1 gives a selection of possible moment distributions. Any one can be selected for the design. On the basis of the theory of plasticity all solutions will furnish the required carrying capacity. However they will exhibit different behaviors under service conditions.

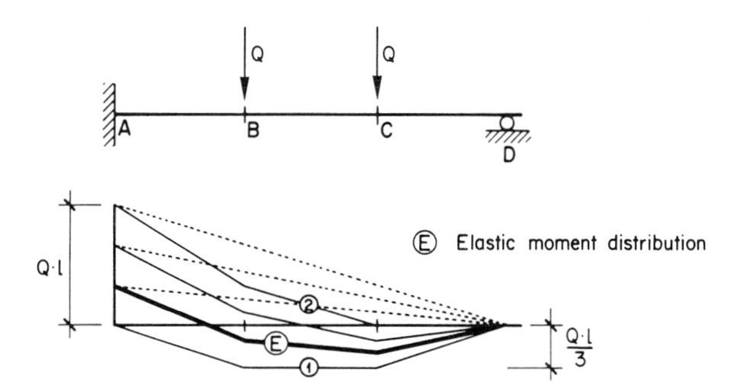

Fig. 4.1: Statically admissible moment distributions

In case the moment distribution ① is used for the design no upper reinforcement is required at the fixed end. However large cracks will develop at this end which cannot be tolerated. Similarly a beam designed on the basis of the moment distribution ② may perform unsatisfactory. In this case the beam is subjected to negative moments only and hence no lower reinforcement is required. This in turn will lead to unacceptable cracks within the span and finally a discontinuity at the right support. In general it can be stated that designs based on extreme moment distributions will produce plastic deformations and unacceptable cracks under service loads. It is even possible that in such structures the required plastic deformations may exceed the available ductility. In such cases the resistance will be smaller than the ultimate load derived on the basis of the theory of plasticity.

Of all possible moment distributions only one distribution fulfils the condition that under a proportional load increase plastic deformations initiate simultaneously at the fixed end and within the span, and hence the ultimate load is reached without a redistribution of the internal forces.

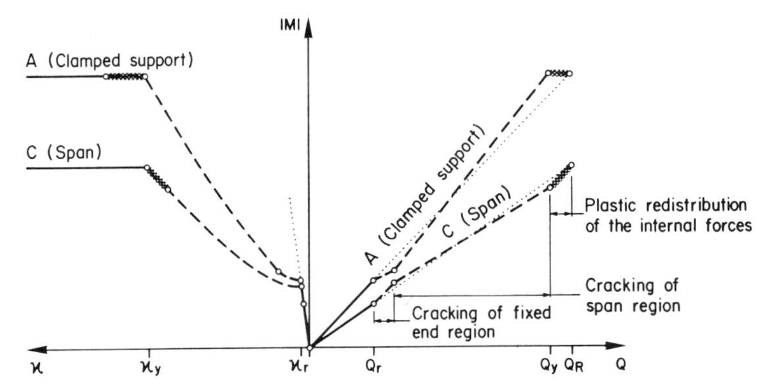

Fig. 4.2: Elastically designed beam, moments as a function of the load [ref. 19]

A linear elastic analysis is based on the assumption that the stiffness of all elements is independent of the load. This assumption holds for reinforced concrete beams only in the uncracked stage (compare Fig. 1.4). Hence even in beams designed elastically a redistribution of the internal forces will occur before the ultimate load is reached. This behavior is illustrated in Fig. 4.2 [ref. 19]. The moment M_A at the fixed end and the moment M_C in the span are shown as a function of the load and the curvature. After reaching the cracking load Q_r the stiffness in the region of the fixed end diminishes. Hence under increasing load the fixed end moment increases at a lower rate than the span moment. Afterwards cracks will develop within the span producing a decrease of the stiffness in the corresponding region. This in turn will lead to an increased growth of the fixed end moment. As the reinforcement at the fixed end and hence the stiffness of the cracked section is greater than within the span, the yield strength of the reinforcement at the fixed end will be reached first. The corresponding load is the yield load Q_y which is smaller than the ultimate load Q_R. The load can be further increased until the plastic moment in the span is reached. This phase is termed redistribution of internal forces.

It follows that it is possible to design a reinforced concrete structure plastically in such a way that its performance under service load is superior to the performance of the same structure designed elastically. The requirements concerning ductility will be less stringent for this plastically designed structure as the redistribution and hence the plastic deformations will be smaller.

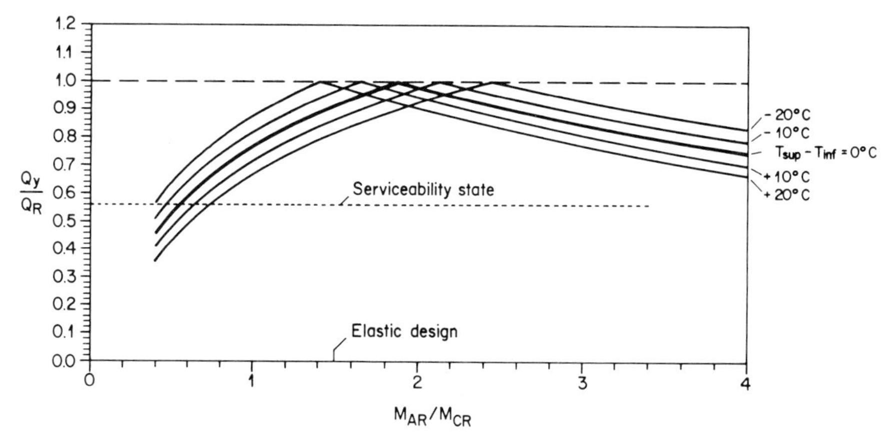

Fig. 4.3: Ratio yield load / ultimate load as a function of the moment ratio M_{AR}/M_{CR} and a residual stress state (linear temperature gradient)

It should however be mentioned, that searching for this optimum solution has no practical significance. The distribution of the internal forces under service load is strongly influenced by residual stresses due to temperature, creep and shrinkage of the concrete, support movements, etc. On the contrary, the plastic ultimate load is independent of such influences. Furthermore, a restricted redistribution of the internal forces before reaching the ultimate load does not essentially influence the behavior under service load. This is illustrated in Fig. 4.3 where the relation between the yield load Q_y, the load initiating the first plastic deformation, and the ultimate load Q_R as a function of the assumed moment ratio, fixed end moment M_{AR} to span moment M_{CR}, for the plastic design is shown. As a residual stress state a linear temperature gradient over the depth of the beam has been assumed.

In conclusion it should be stated that in the design of a reinforced concrete structure on the basis of the theory of plasticity, a considerable free margin in the selection of the distribution of the internal forces exists. This margin can be fully utilized without causing adverse effects in the behavior under service loads.

4.1.2 Selection of the Inclination of the Compression Field in the Web of Beams

Similar considerations can be made for the selection of the inclination of the compression field in the web of girders. In section 2.3.1 it was shown that a decrease of the inclination leads to a decrease of the required stirrup reinforcement and an increase of the horizontal tension force in the support region. As in the previously investigated case where extreme distributions of the reinforcements should be avoided (Fig. 4.4a), extreme angles of inclination of the compression field should not be selected (Fig. 4.4b). For an angle ϑ too small a correspondingly small stirrup reinforcement results from design such that excessive cracking will already develop in the web under service loads. For an angle too large a small tension force results from design such that the anchorage of the tension force in the support region can cause problems under service loads.

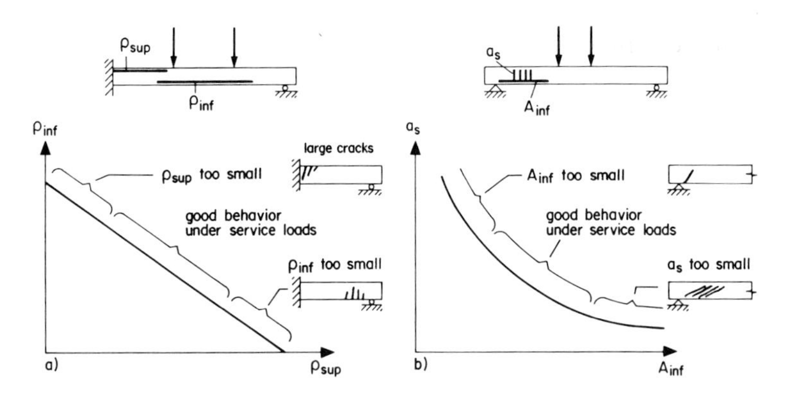

Fig. 4.4: Influence of the distribution of the reinforcement on the behavior under service loads

Nevertheless, a considerable freedom in the selection of the inclination of the compression field is available. It should also be mentioned that the exact determination of the stress fields occurring under service and ultimate loads is of little importance. They are fixed by the selection of the reinforcements, hence by the design, and not in the reversed order.

4.1.3 Redistribution of Internal Forces and Ductility Requirements

For extremely brittle material behavior the collapse load can be overestimated using both the elastic and the plastic design procedures. The greater the required plastic redistribution of the internal forces, the greater is the reduction of the collapse load due to insufficient ductility in such cases.

As mentioned above, in general, both for elastically and plastically designed structures, a redistribution of the internal forces is necessary before reaching the collapse load. According to Fig. 4.3, in plastic design there exists considerable freedom for the choice of the ratio of the resistances, whereby the required redistribution of the internal forces is of the same order of magnitude as for an elastically designed structure. It follows that the ductility requirements of plastically designed structures are comparable to those of an elastically designed structure.

Further, it should be mentioned that in both design methods the basic principle remains valid that the structural components have to be detailed such that they exhibit the maximum possible ductile behavior. Only in this way, irrespective of the design method employed, a satisfactory behavior of the whole structure can be achieved for extraordinary load cases.

4.2 PROCEDURE FOR DEVELOPING STRESS FIELDS

4.2.1 Introduction

Statically indeterminate systems can be conceived as a combination of statically determinate systems. For beam and frame structures possible sub-systems can be identified readily. The system of Fig. 4.1 is statically indeterminate to the first degree such that a combination of two sub-systems is sufficient. It can be subdivided into a simply supported beam and a cantilever beam.

Walls, slabs and three-dimensional bodies are internally highly indeterminate systems. For their design it is necessary to identify the relevant sub-systems. Despite the fact that theoretically an infinite number of sub-systems are eligible, it is sufficient to select a limited number of relevant sub-systems in order to achieve adequate behavior under service as well as under ultimate loads. In this section the selection of these relevant sub-systems will be illustrated by making use of kinematic considerations.

As an introductory example the wall with an opening, shown in Fig. 4.5, will be treated. The procedure for developing an adequate stress field is illustrated in Fig. 4.6.

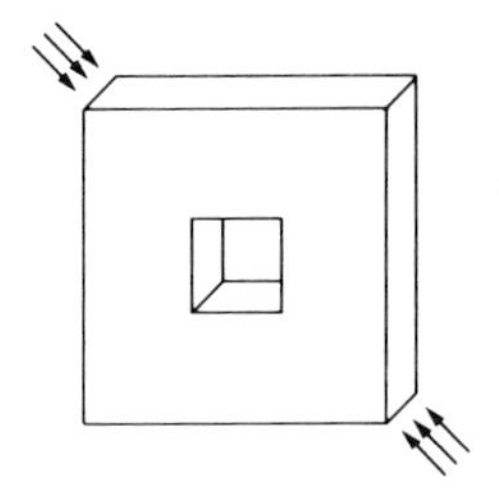

Fig. 4.5: Reinforced concrete wall element and loading

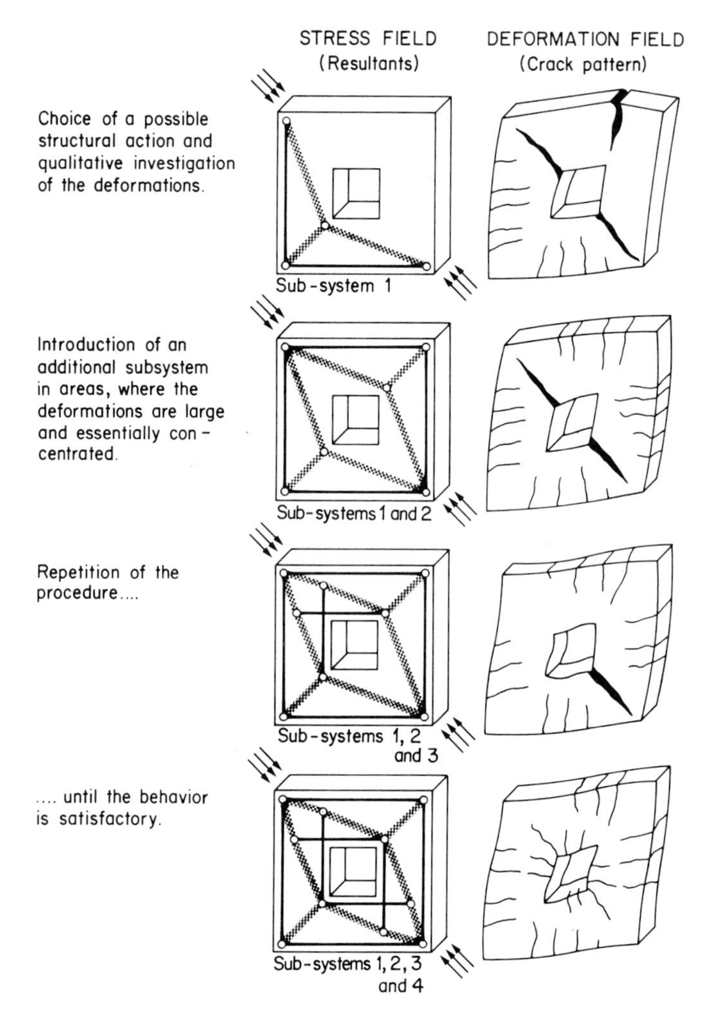

Fig. 4.6: Procedure for developing suitable stress fields

Firstly, a possible structural action is freely chosen. In the next step the associated deformations are investigated qualitatively. Thereby, it is assumed that in the unreinforced parts the deformations are essentially concentrated in open cracks, so the corresponding elements may be considered to be rigid-plastic. In order to avoid these considerable concentrated deformations, an additional sub-system is introduced in these areas.

This procedure is repeated until a satisfactory deformation field is obtained. In the example presented the four sub-systems are introduced in steps for didactic reasons. By considering symmetry in the first place two of the steps could have been eliminated.

4.2.2 Spreading of Force in a Wall Element Loaded in Tension

As a first sub-system a direct force transmission with a tension band along the middle as shown in Fig. 4.7a is selected. The associated type of deformation (crack pattern) is characterized by a distribution of cracks in the reinforced region only (Fig. 4.7b). In order to avoid the formation of only a few large cracks in the outer regions an additional sub-system has to be introduced. This is characterized by a force transmission to these outer regions as well (indirect force transmission, Figs. 4.7c and d).

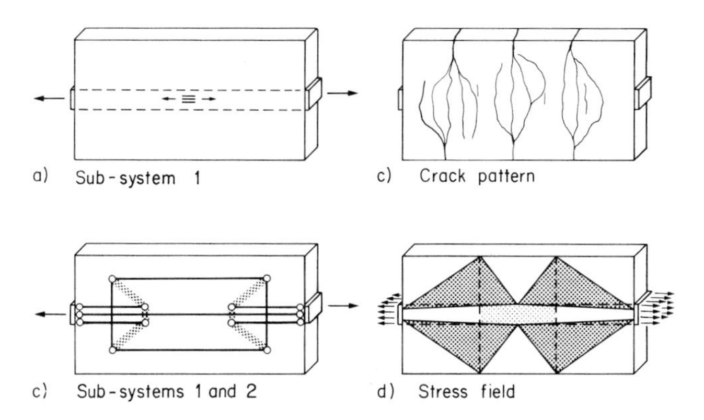

Fig. 4.7: Wall element loaded in tension

Since the tension bars of the sub-system 2 are not stretched as much as those in sub-system 1, the boundary regions can be provided with less reinforcement. In general, it suffices in this region to place minimal reinforcement which is just able to distribute the cracks.

4.2.3 Spreading of Force in a Wall Element Loaded in Compression

As in the case of the element loaded in tension the load can be directly transmitted in a small compression band (Fig. 4.8a). Such a stress field does not require reinforcement. A crack pattern with a crack opening in the axial direction (tensile split) can develop. The element is almost separated into two parts, which are loaded eccentrically. The eccentric loading leads to a curvature of the two parts. As a result the crack opening is large. The two separated parts can be held together with the aid of transverse reinforcement. A proportion of the axial compressive force can spread outwards and is deviated by means of this reinforcement (Fig. 4.8b).

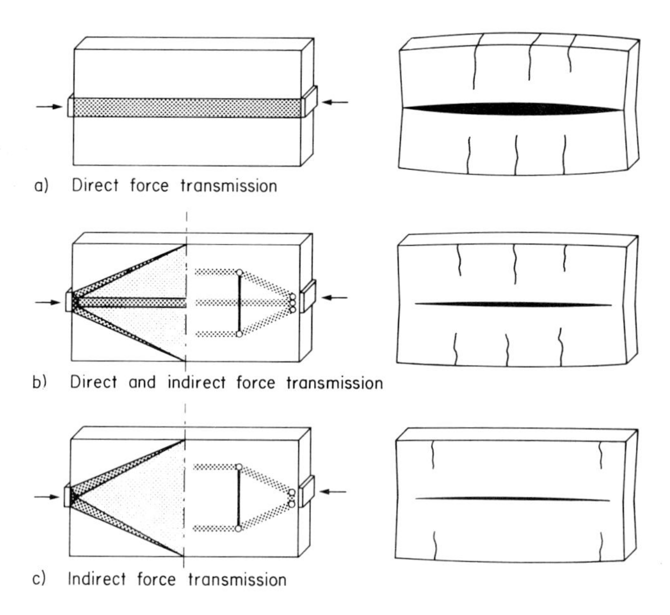

Fig. 4.8: Wall element loaded in compression

The opening of the axial crack depends on the transverse reinforcement. If as a consequence of increasing the amount of transverse reinforcement the whole force is deviated (no more direct force transmission) then only a small crack opens (Fig. 4.8c). The sub-system 1 (direct force transmission) practically disappears in this case.

It should be observed that the transverse reinforcement should be placed where the spreading of the force accurs. The extension of this zone is a function of the amount of reinforcement and the geometry of the element. Thus the requirements of serviceability can be fulfilled without any problem.

As mentioned already in chapter 3, a wide crack can also be problematic with respect to strength. Fig. 4.9 shows an element without any transverse reinforcement loaded in compression which exhibits a wide longitudinal crack in the middle of the element. Since this crack does not always run parallel to the compression strut a part of the force must be carried by interlocking action at the crack boundary.

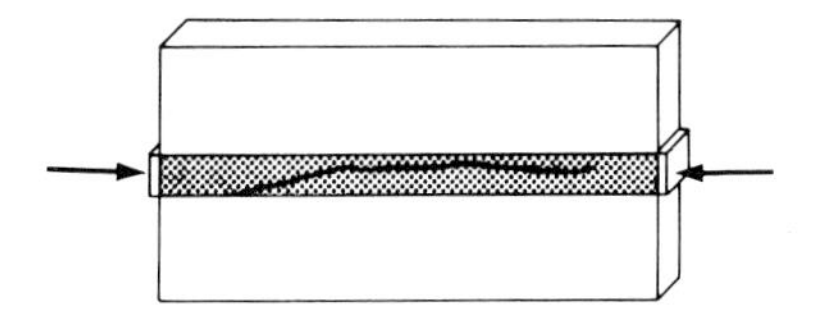

Fig. 4.9: Influence of wide crack on strength

To obtain the envisioned strength it is necessary to provide local reinforcement ties for the compression strut. By having a transverse reinforcement over the whole width of the element both serviceability and strength are achieved.

The two examples given in this section exhibit important differences with regard to keeping the crack small:

- In the case of a tension element, for a given longitudinal reinforcement the total crack width is more or less given; the cracks have to be distributed.

- In the case of a compression element, on the other hand, the total crack width can be influenced by the transverse reinforcement.

4.2.4 Further Cases

The spreading of stress fields presented in the foregoing section may be applied analogously to all design cases in which the first preliminary assumption of a stress field results in concentrated struts.

Fig. 4.10 shows several applications with a preliminary assumption of stress fields, the expected crack pattern and the final stress field on which the design is based. It is to be remarked, that in the design of such elements it is often sufficient to study the qualitative development of the stress field. The required reinforcement to fulfil the conditions of serviceability can be obtained by adequate estimates.

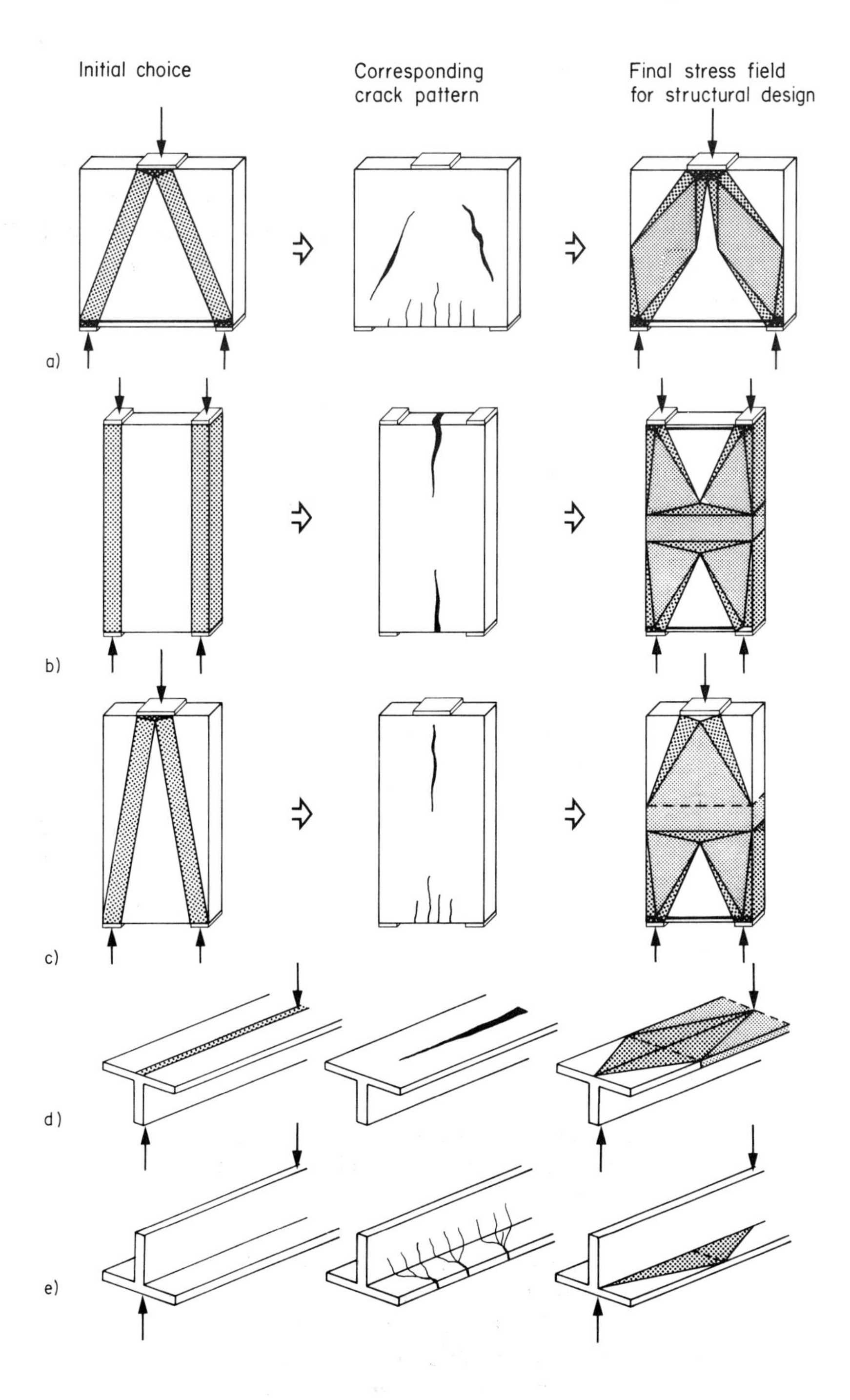

Fig. 4.10: Crack control by spreading of stress fields

4.3 IMPAIRED STRENGTH THROUGH WIDE CRACKS

In chapter 3 it was shown that a complete force transfer by means of interlocking action at crack boundaries is only possible if the crack width is kept small. In extreme cases for reinforced concrete elements without distributed reinforcement it may happen that cracks that are too wide occur in a region in which a compression strut should develop. Such cracks may arise from

- previous structural actions
- previous load cases
- residual stress states (temperature, shrinkage, settlements, etc.).

The redistribution of internal forces is impaired in such cases, so that premature failure may occur. Problems then arise when a crack intersects the strut at a shallow angle and the force has to be transferred by means of interlocking action. Such a case has been dealt with in section 4.2.3 (Fig. 4.9). A further example for the reduction of strength due to wide crack openings is shown in Fig. 4.11.

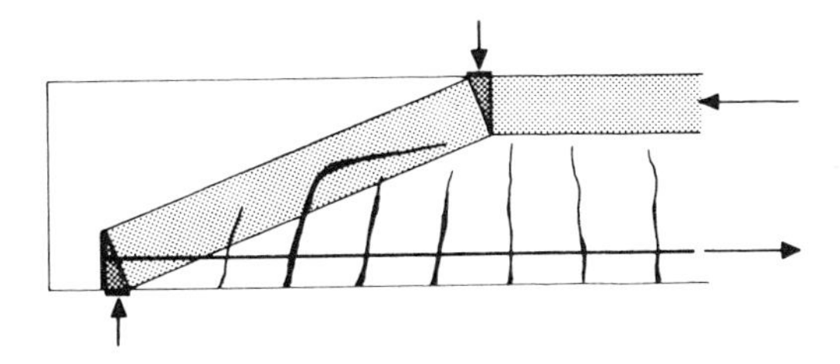

Fig. 4.11: Crack pattern and stress field

When the load is increased in this case several structural actions can develop. Firstly, more or less vertical cracks due to bending action are formed, which during a further load increase develop into horizontal cracks. Measurements carried out on test specimens show that there can be no complete force transfer over these cracks [19]. Therefore the strut shown in Fig. 4.11 cannot develop. As discussed in chapter 2 for such beams the load is usually carried indirectly by stirrups. A direct force transfer is only possible in this case if the cracks in the strut zone are kept small by means of local ties.

The following conclusion can be drawn from these two examples:

> *In order to avoid a loss in static strength due to cracking, compression struts without stirrup reinforcement over longer distances should be avoided. The situation becomes critical when the angle between a compression strut and a tension tie is small.*

4.4 STRESS DISTRIBUTION IN HIGHLY STRESSED COMPRESSION ZONES

4.4.1 Introduction

For the stress fields developed in chapter 2 a constant stress over the depth of the compression strut has been assumed. With regard to ductility requirements this assumption is only valid if the depth of the strut is small compared to the depth of the section. In such cases a varying stress distribution, as shown e.g. in Fig. 2.5b, leads to an insignificant displacement of the compression resultant and hence a negligible influence on the strength. In this section the influence in the case of a compression strut with a relatively large depth is investigated.

4.4.2 Stress Fields for Beam-Columns

A beam-column is a compression member subjected to transverse bending and axial compression. First the case of relatively small eccentricities and hence small bending moments and shear forces is investigated (Fig. 4.12a).

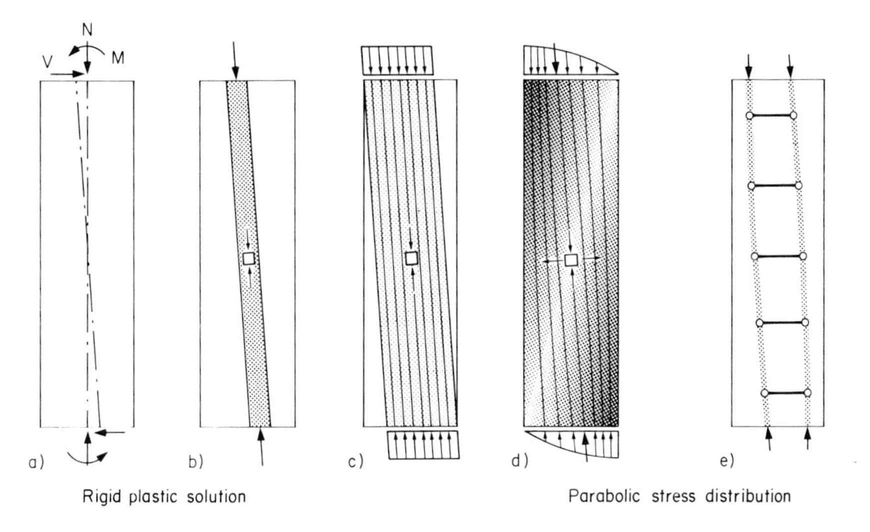

Fig. 4.12: Beam-column with small eccentricity of the normal force

Two rigid-plastic solutions are given, one with a stress level f_{ce} in Fig. 4.12b, and the other with the largest possible width of the compression strut and a correspondingly smaller stress level in Fig. 4.12c.

A solution with a parabolic stress distribution is given in Figs. 4.12d and e. The stress trajectories of the compression strut are curved such that transverse tensile forces develop. The latter have to be resisted by tensile stresses in the concrete or in the case of cracking by transverse reinforcement. As the cracks develop parallel to the compression field the strength will not be impaired. As no tensile stresses are transferred across the cracks, a redistribution of the internal forces will take place leading to a stress field as shown in Fig. 4.12c. This field can therefore be used for the design.

For a given normal force and a given concrete strength f_{ce} the assumption of a constant stress level f_{ce} in the compression strut furnishes the largest possible eccentricity of the resultant. Of particular interest is the case of the maximum eccentricity and hence the maximum bending moment which can be resisted without any tensile reinforcement. If however for this limiting case a nonlinear stress distribution with a strain limit (Fig. 4.13b) is assumed, tensile reinforcement will be necessary to resist the said moment.

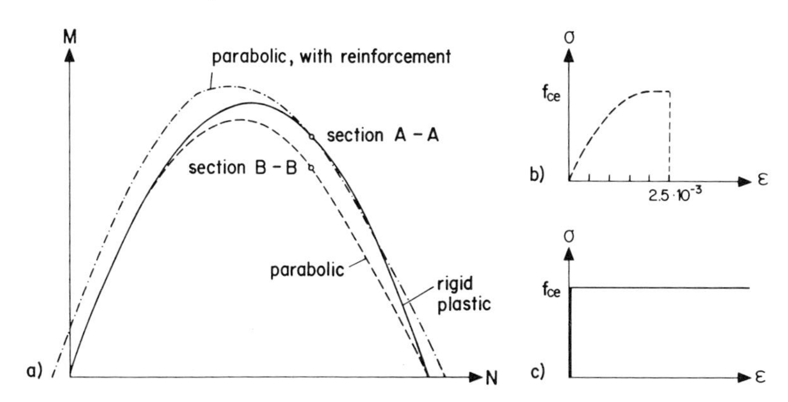

Fig 4.13: *M-N* Interaction diagram in dependence of the material behavior

The corresponding stress fields are shown in Fig. 4.14. For the design of the longitudinal reinforcement the end section *A-A* is governing. The corresponding *M-N* Interaction diagram is given in Fig. 4.13a.

Longitudinal reinforcement is required up to section *B-B*. The *M-N* Interaction diagram shows that beyond this section the bending moment is resisted solely by the eccentricity of the compression resultant. Using simple equilibrium considerations the part of the shear force to be resisted indirectly, i.e. by action of stirrups, can then be determined. For this design the stress field of Fig. 4.14e can be used. Here a constant stress distribution with a stress level smaller than f_{ce} has been chosen. However the same locations of the resultants in section *A-A* and *B-B* as determined for the stress fields of Fig. 4.14c are retained.

It becomes evident that beam-columns subjected to intensive normal forces require particular attention for the region with an intermediate eccentricity of the axial force. Besides the longitudinal reinforcement a stirrup reinforcement is statically required.

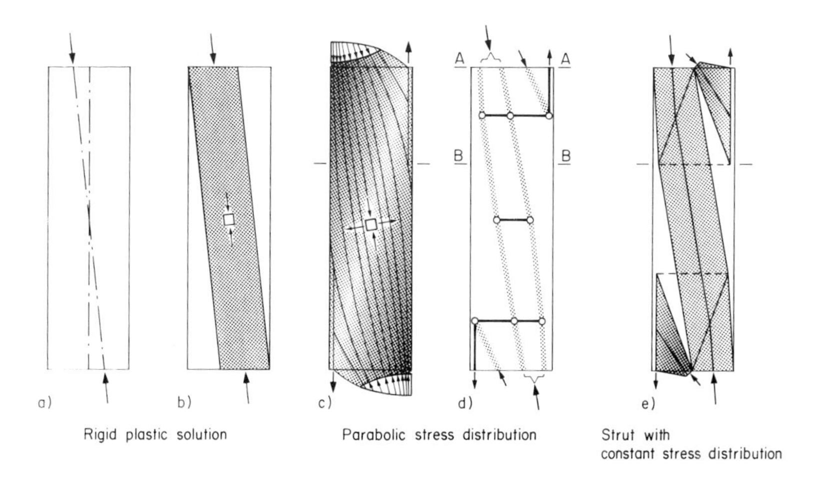

Fig. 4.14: Beam-column with intermediate eccentricity of the normal force

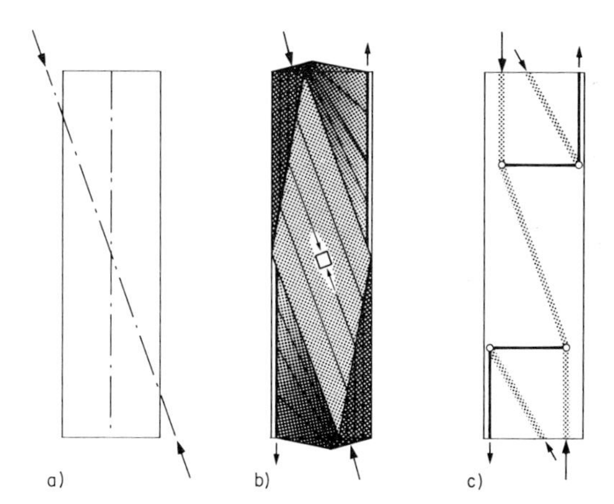

Fig. 4.15: Beam-column with large eccentricity of the normal force

For beam-columns with a large eccentricity of the normal force similar design considerations hold. Concerning the choice of the inclination of the compression field the observations made for beams, sections 2.3.1 and 4.1.2, should be considered.

4.5 PRESTRESSED BEAMS

4.5.1 Beam with straight cable

Since the prestressing only causes a residual stress state in the element it has no influence on the load bearing capacity determined on the basis of the theory of plasticity. However, since smaller or greater redistributions of the internal forces are necessary in order to reach the limit state, depending on the stress field, it is reasonable in many cases to introduce the prestressing forces as external forces acting on the beam.

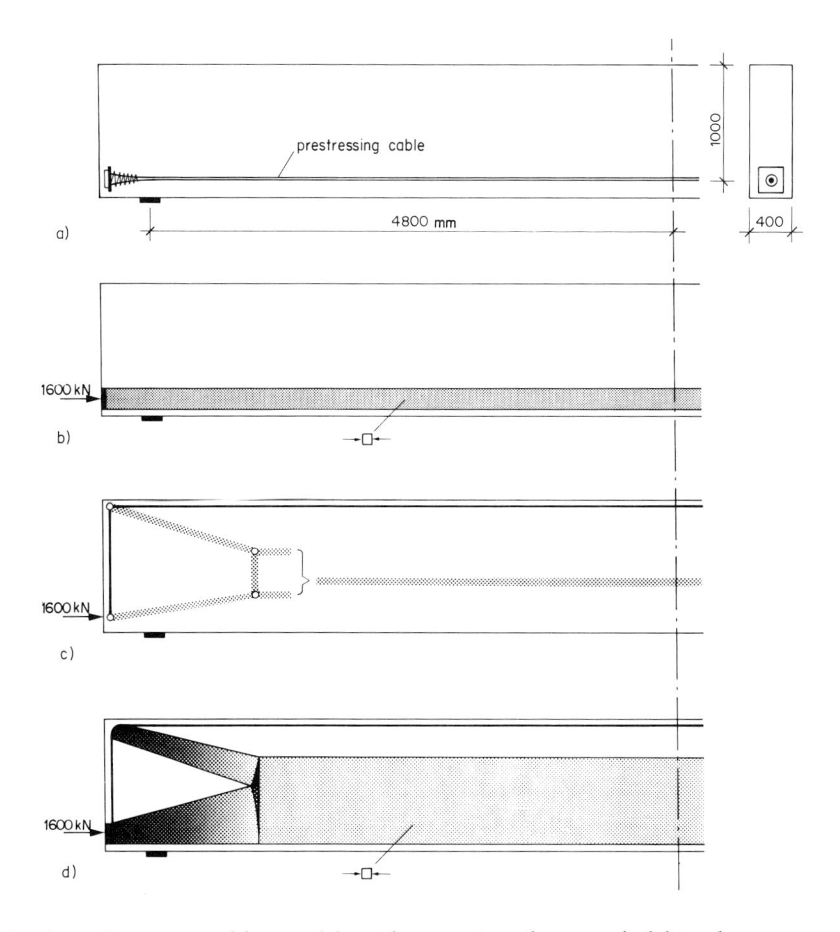

Fig. 4.16: Prestressed beam (a) with associated stress field in the concrete (b) and stress resultants (c) with stress field (d) taking into account the force transfer mechanism

In this way it is guaranteed that the stress field takes into account the introduction of the prestressing force. Besides the anchor, deviation and friction forces the additional forces resisted by the prestressing cables result only from increases in the external loads, so that the prestressing cables can be treated as passive reinforcing bars of nominal tensile strength (f_{py}-σ_p), in which σ_p is the stress in the cables due to prestressing only.

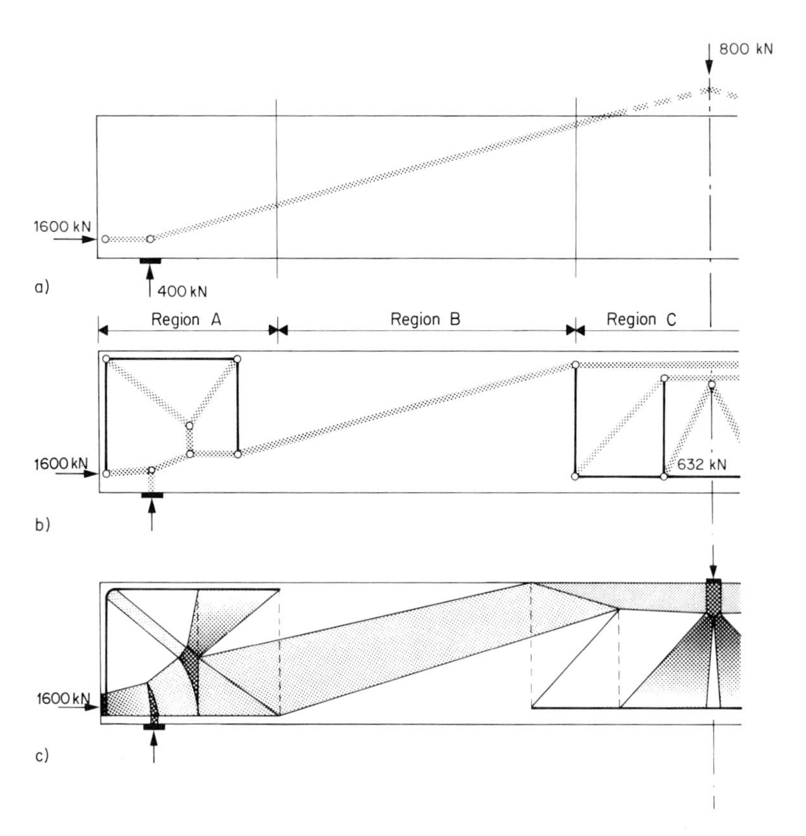

Fig. 4.17: Prestressed beam with loading

Due to the anchor forces the beams thus modeled are usually loaded in compression, so that similar stress fields to those developed for the columns in section 4.4 can be used. In Figs. 4.16a and b the structural action of an beam due to prestress only is shown. According to section 4.2.3 the spreading out of the compressive forces in the anchor zone has to be considered as shown in Figs. 4.16c and d.

Analogous to the case of columns the external force normal to the axis of the beam produces an inclination of the global stress resultants. Force transfer has to be considered in this case as well. The structural action is evident from Fig. 4.17. In the anchor zone with a large eccentricity of the global stress resultant (region *A*) the variation of the internal forces is analogous to that of the unloaded beam, Fig. 4.16c. The splitting force concentrated at the end of the beam, however, is reduced due to the shear force.

In region *C*, where the global compressive stress resultant lies partly outside the beam, an increase of the force is necessary in the prestressing cables or in any passive reinforcing bars. In this region the structural action is identical with that of a beam with passive reinforcement and the same considerations for the choice of the compression diagonal apply as described in section 4.1.4.

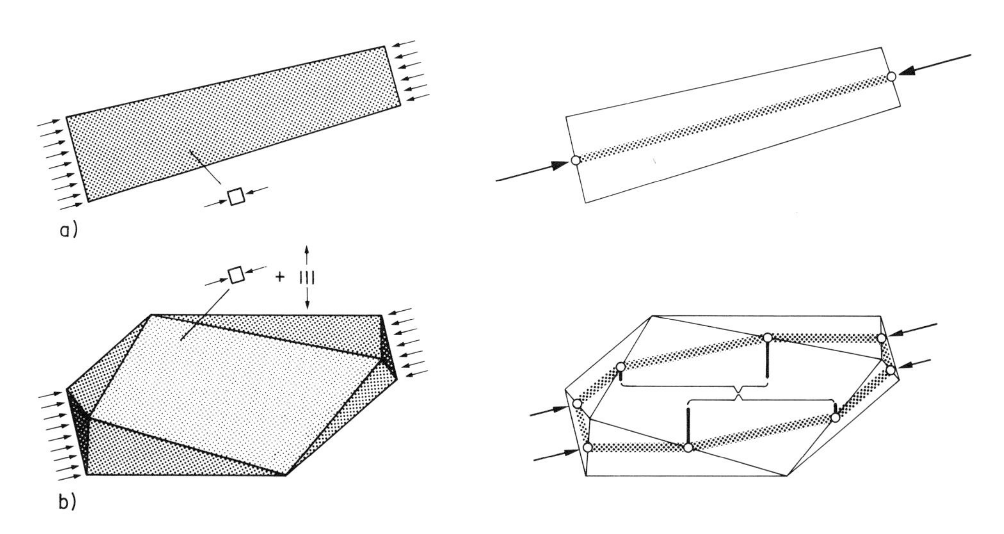

Fig. 4.18: Spreading of the inclined strut in region *B*, Fig. 4.17

In region B, in which the eccentricity of the compression resultant is small, the force in the prestressing reinforcement remains essentially unchanged by the external load. Thus the dangerous cracks shown in Fig. 4.11 cannot develop. If there are no wide cracks resulting from constraints or other load cases present then a direct compression strut can develop in this zone.

Since cracks can be kept small by having just minimum stirrup reinforcement it is hence not necessary to develop a detailed stress field in region *B*. For the sake of completeness, however, this stress field, which replaces an inclined strut, is shown in Fig. 4.18b. The small minimum stirrup reinforcement leads to a small inclination of the strut, which can develop without any problem.

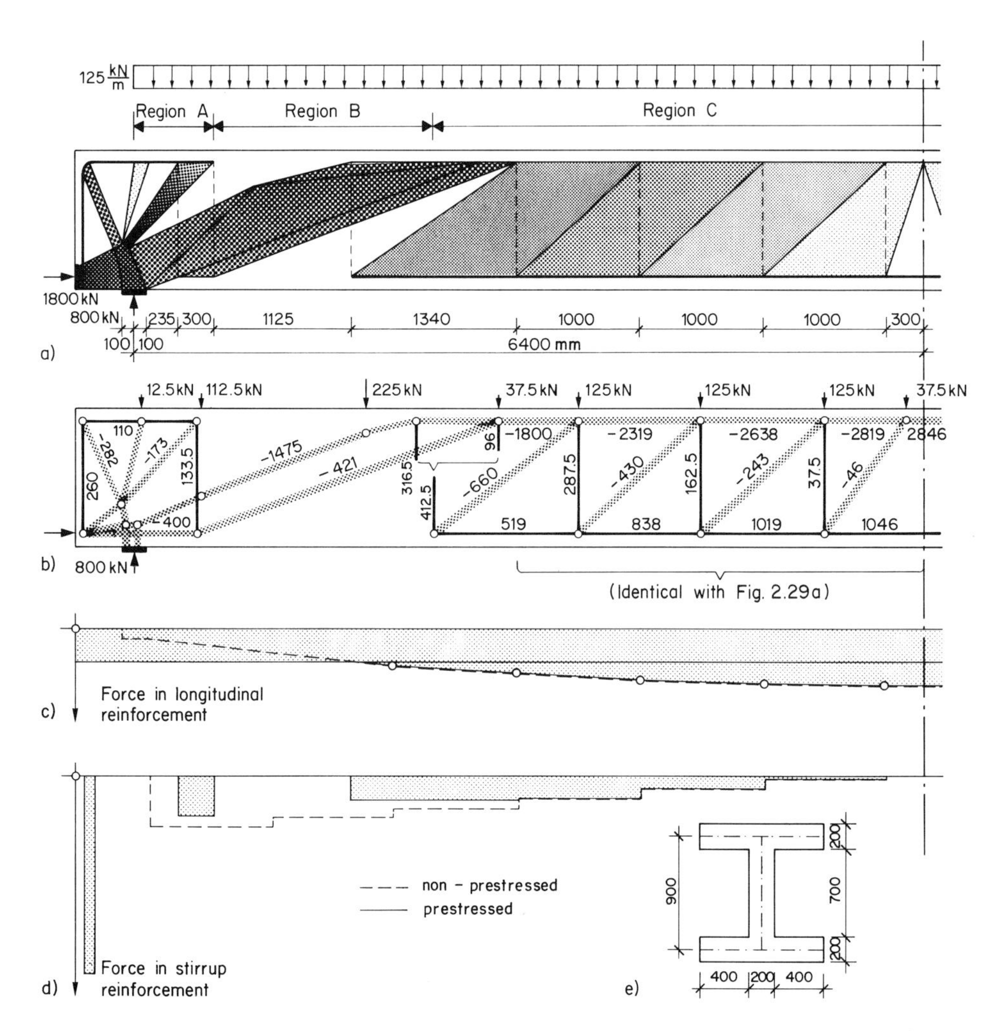

Fig. 4.19: Prestressed beam under distributed load, comparison of the forces in the reinforcement with those of a beam with passive reinforcement

Beams subjected to other types of loading and other forms of cross-section may be treated in the same way. Fig. 4.19 shows the stress field of the beam with an I-section under distributed load treated in section 2.3.2. By way of comparison the distributions for the longitudinal and stirrup forces for the beams with and without prestressing are also shown.

4.5.2 Beam with curved cable

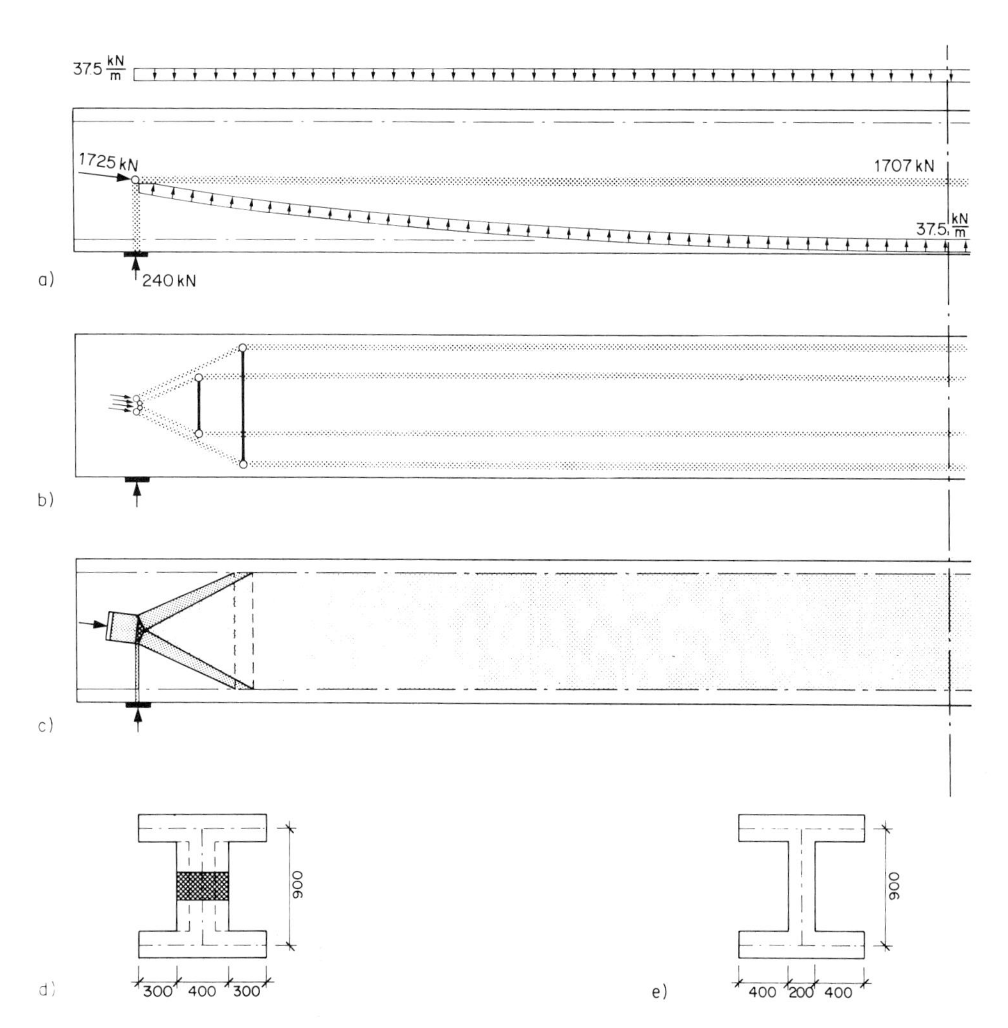

Fig. 4.20: Beam with curved cables, special case (deviation forces balancing dead load)

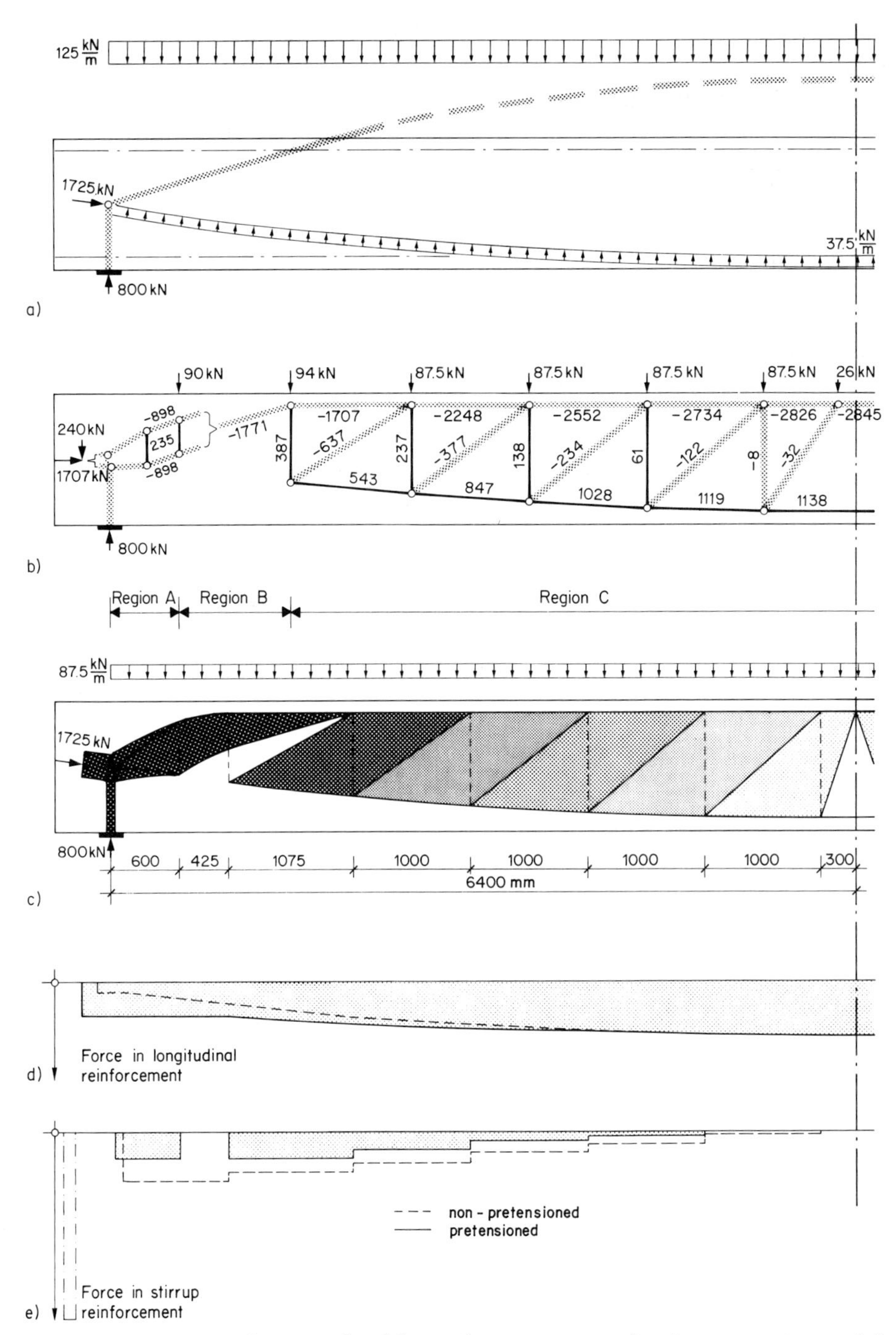

Fig. 4.21: Beam with curved cable under maximum load, comparison of the forces in the reinforcement with those of the non-pretensioned beam

If the cables are curved their effects on the beam due to the deviation forces must also be considered. It should be noted that in the support region an axially acting anchor force spreads out differently from that of a strongly eccentric anchor force. The size of the splitting force is a maximum under a load configuration in which the external loads are balanced by the deviation forces and hence the global stress resultant is horizontal as shown in Fig. 4.20.

For the dimensioning of the stirrup reinforcement in region *C*, the load case with the maximum load is governing. The structural action is characterized by the variable static height (Fig. 4.21). The slope of the cable causes an unloading of the stresses in the web, so that the required stirrup force is reduced (Fig. 4.21c). The internal forces produced by the deviation forces are not pursued in further detail. They compensate directly a part of the external loading.

Often, besides the prestressed cable reinforcement, horizontal passive reinforcement is also present. Fig. 4.22 shows how this reinforcement together with the cables is activated in region *C*.

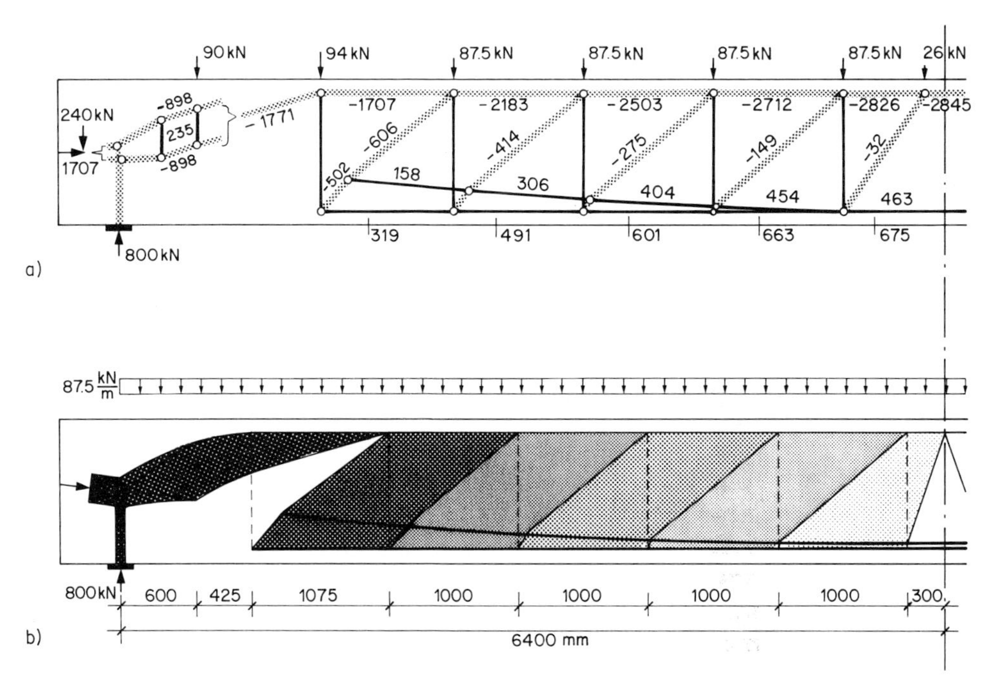

Fig. 4.22: Beam with additional passive reinforcement

4.5.3 Anchorage zone of pretensioned beams

In the case of pretensioned prestressed beams the prestressing force is transferred to the beam by bond. Based on the stress field shown in Fig. 4.23a it is evident that the splitting reinforcement has to be distributed.

If the support reaction acts in a concentrated manner at the end of the beam (Fig. 4.23b) then special attention has to be given to this region. In general the prestressing wires cannot be sufficiently anchored in the support region. Hence the required tensile force has to be resisted by adequate passive reinforcement and transferred to the prestressing wires. It should be noted that the stirrup force required in this region results from the superposition of the shear force and the splitting force.

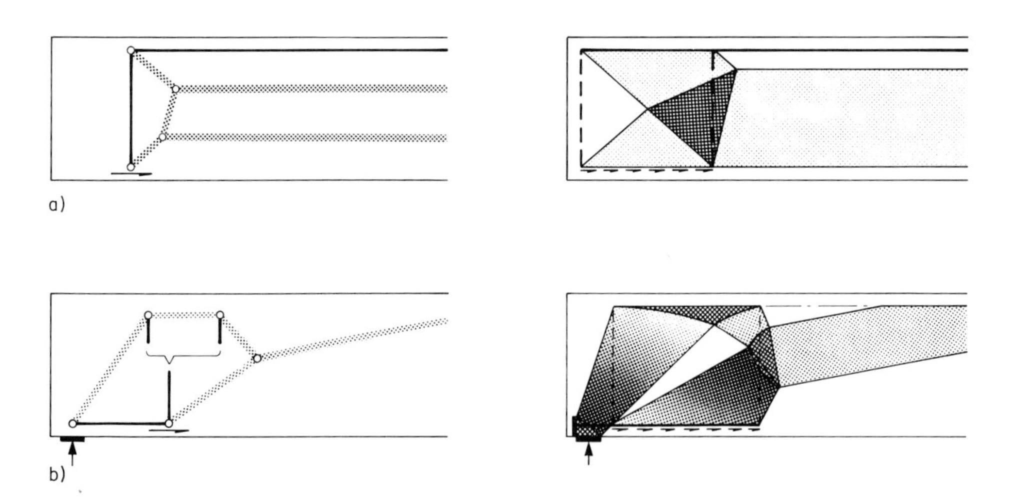

Fig. 4.23: Structural action at the end of a pretensioned beam with anchorage of wires by bond; global stress resultants due to (a) prestress only and (b) prestress and support reaction

4.5.4 Unbonded prestressed beams

In the case of prestressing cables without bond a similar stress field as shown in Fig. 4.22 can be developed. There is no deviation of the diagonal compression field in the support region. The cable force remains constant and there is only an increase of the force in the passive reinforcement.

5 PLANE STRESS, PLATE AND SHELL ELEMENTS

In chapter 2 it was shown that for the dimensioning of beams basically two methods are possible. The first method, called the *Integral Method,* involves developing a stress field for the whole structure. In the second method, called the *Sectional Method,* the stress resultants are determined firstly using conventional statical methods (elastic and plastic analysis). The dimensioning is then carried out for the governing regions with the aid of equilibrium considerations in suitable sections. In the case of simple sectional forms and distributed loads design formulas have been derived (Figs. 2.36, 2.51 and 2.59).

From chapter 2 it is evident that the *Integral Method* presents an efficient procedure of designing elements subjected to plane stress. In this case the distribution of the internal forces is investigated to determine directly the required amount of reinforcement and to estimate the stresses in the concrete. Thanks to modern computational methods (finite element programs, etc.) the *Sectional Method* has gained in importance. The sectional forces or stress resultants (n_x, n_y, n_{xy}) are determined and from these the design quantities (forces in the reinforcement and the concrete) are derived. This step is carried out by means of the design equations developed in section 5.1.

In the design of slabs, only the *Sectional Method* is of practical significance. The stress resultants (m_x, m_y, m_{xy}) can be determined using elastic or plastic analysis. In the first case mainly numerical methods are employed (finite element programs). According to the theory of plasticity refined and less refined equilibrium solutions (moment fields) are available. Since they are not further pursued here interested readers are referred to the publication by Hillerborg [34] and Marti [35]. In the following the design equations which define the design quantities as a function of the stress resultants are derived (section 5.2).

In the case of shells the same considerations apply as for plane stress and slab elements, depending on whether the membrane action or the bending action is governing. The design equations are derived in section 5.3.

5.1 PLANE STRESS ELEMENTS

The element shown in Fig. 5.1 is subjected to the forces n_x, n_y and n_{xy}. The design involves determining the amount of reinforcement for the given directions of the reinforcing bars and choosing the thickness of the element such that the effective concrete strength is not exceeded.

The trivial case in which both principal stresses are negative (compressive), thus requiring no reinforcement, needs no further consideration.

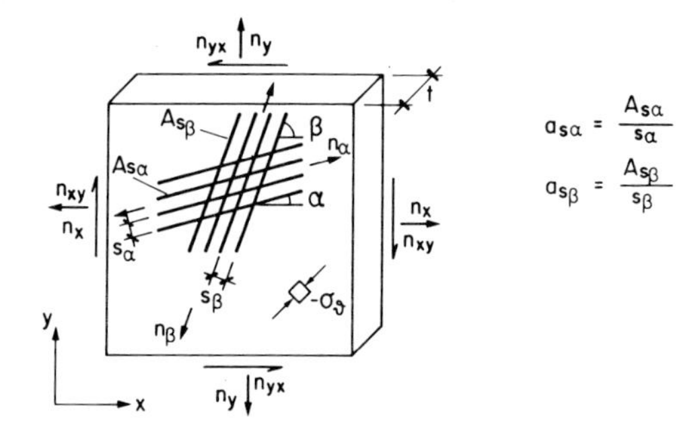

Fig. 5.1: Plane stress element

The steel forces $n_\alpha = \sigma_{s\alpha} \cdot a_{s\alpha}$ and $n_\beta = \sigma_{s\beta} \cdot a_{s\beta}$ together with the force in the concrete $n_\vartheta = \sigma_\vartheta \cdot t$ are found separately for each stress resultant. Fig. 5.2a shows a plot of the internal forces due to n_x. It should be noted that the inclination ϑ of the compression field may be chosen as for the web of a beam (section 2.3.1 and 4.2).

The equilibrium conditions can be formulated in a simple manner if sections are placed parallel to the direction of the reinforcement. By a suitable adjustment of the element to the given reinforcement directions as shown in Fig. 5.3 the resultants of the stress field in a plane stress element can be presented in a simple manner.

The resulting forces in the steel and concrete are calculated using the force diagram (Fig. 5.2c). The following values per unit width are obtained:

$$n_\alpha = \frac{n_x}{\cos^2\alpha} \cdot \frac{\tan\beta \cdot \tan\vartheta}{(\tan\alpha - \tan\beta) \cdot (\tan\alpha - \tan\vartheta)} \qquad (5.1)$$

$$n_\beta = \frac{n_x}{\cos^2\beta} \cdot \frac{\tan\alpha \cdot \tan\vartheta}{(\tan\beta - \tan\alpha) \cdot (\tan\beta - \tan\vartheta)} \qquad (5.2)$$

$$n_\vartheta = \frac{n_x}{\cos^2\vartheta} \cdot \frac{\tan\alpha \cdot \tan\beta}{(\tan\vartheta - \tan\alpha) \cdot (\tan\vartheta - \tan\beta)} \qquad (5.3)$$

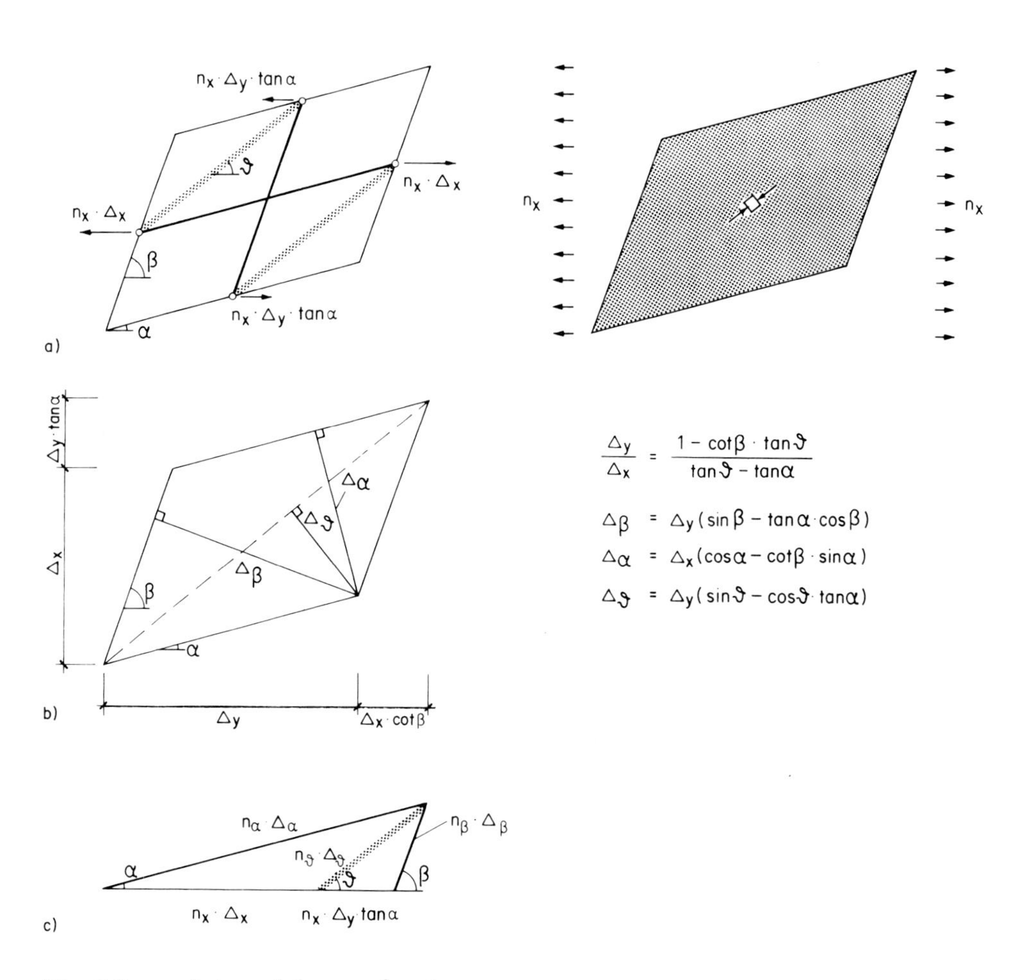

Fig. 5.2: Internal forces due to n_x

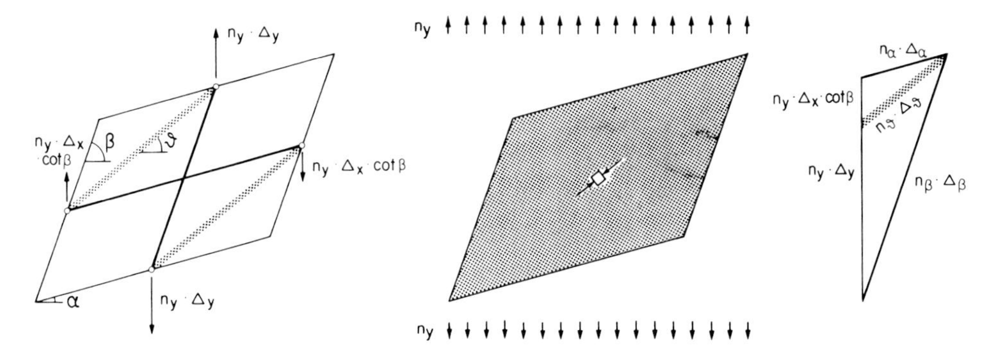

Fig. 5.3: Internal forces due to n_y

For the stress resultant n_y shown in Fig. 5.3 the internal forces are :

$$n_\alpha = \frac{n_y}{\cos^2\alpha} \cdot \frac{1}{(\tan\alpha - \tan\beta) \cdot (\tan\alpha - \tan\vartheta)} \tag{5.4}$$

$$n_\beta = \frac{n_y}{\cos^2\beta} \cdot \frac{1}{(\tan\beta - \tan\alpha) \cdot (\tan\beta - \tan\vartheta)} \tag{5.5}$$

$$n_\vartheta = \frac{n_y}{\cos^2\vartheta} \cdot \frac{1}{(\tan\vartheta - \tan\alpha) \cdot (\tan\vartheta - \tan\beta)} \tag{5.6}$$

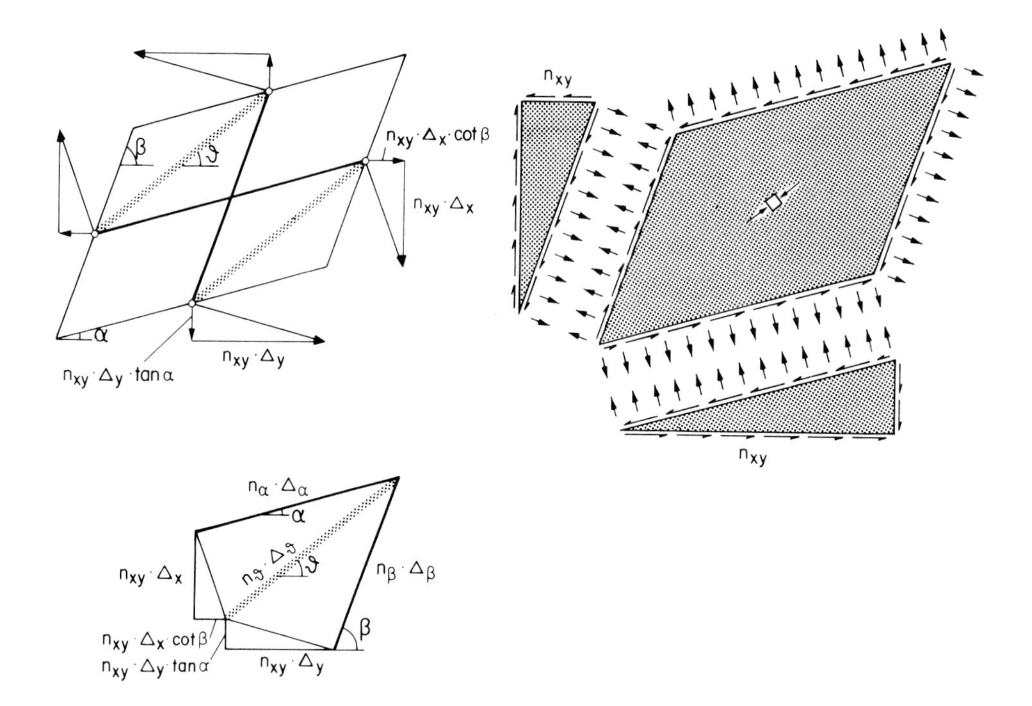

Fig. 5.4: Internal forces due to n_{xy}

Finally the stress resultant n_{xy} (Fig. 5.4) generates the forces:

$$n_\alpha = \frac{n_{xy}}{\cos^2\alpha} \cdot \frac{\tan\beta + \tan\vartheta}{(\tan\alpha - \tan\beta) \cdot (\tan\alpha - \tan\vartheta)} \tag{5.7}$$

$$n_\beta = \frac{n_{xy}}{\cos^2\beta} \cdot \frac{\tan\alpha + \tan\vartheta}{(\tan\beta - \tan\alpha) \cdot (\tan\beta - \tan\vartheta)} \tag{5.8}$$

$$n_\vartheta = \frac{n_{xy}}{\cos^2\vartheta} \cdot \frac{\tan\alpha + \tan\beta}{(\tan\vartheta - \tan\alpha) \cdot (\tan\vartheta - \tan\beta)} \tag{5.9}$$

For the special case of an orthogonal reinforcement in the x and y directions the forces shown in Fig. 5.5 result. Uniaxial loading in the x or y direction is carried directly by the corresponding reinforcement. In the case of loading due to n_{xy} the derived relations simplify to: ($\alpha = 0, \ \beta = \pi/2$)

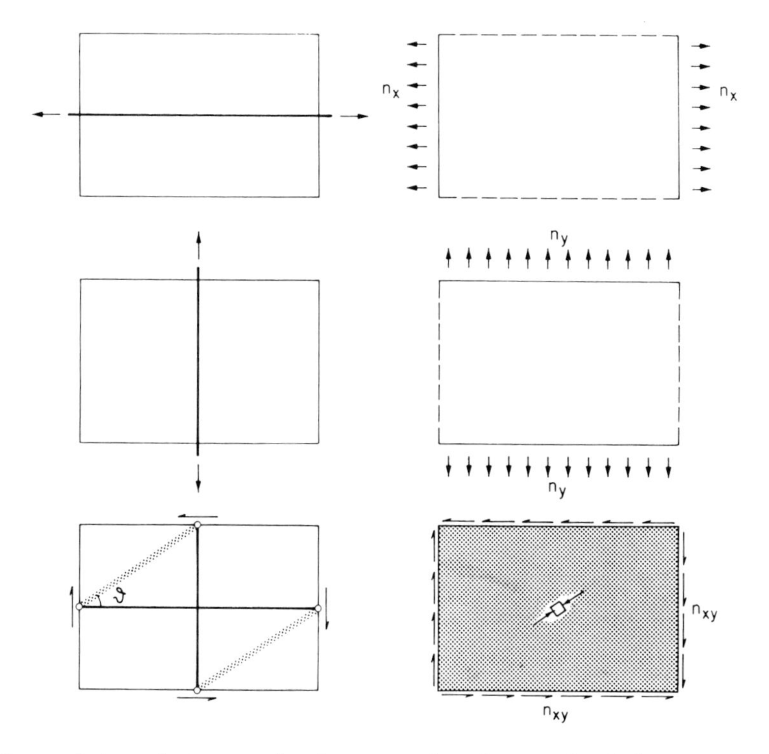

Fig. 5.5: Internal stresses in the case of orthogonal reinforcement

$$n_\alpha = n_{xy} \cdot \cot\vartheta \tag{5.10}$$

$$n_\beta = n_{xy} \cdot \tan\vartheta \tag{5.11}$$

$$n_\vartheta = n_{xy} \cdot \frac{1}{\sin\vartheta \cdot \cos\vartheta} \tag{5.12}$$

If all three stress resultants act simultaneously the design may be carried out as follows:

- The inclination ϑ of the compression field is selected. The resultant forces n_α, n_β and n_ϑ are obtained by superposition of the values due to n_x, n_y and n_{xy}:

$$n_\alpha = \frac{n_x \cdot \tan\beta \cdot \tan\vartheta + n_y + n_{xy} \cdot (\tan\beta + \tan\vartheta)}{\cos^2\alpha \cdot (\tan\alpha - \tan\beta) \cdot (\tan\alpha - \tan\vartheta)} \tag{5.13}$$

$$n_\beta = \frac{n_x \cdot \tan\alpha \cdot \tan\vartheta + n_y + n_{xy} \cdot (\tan\alpha + \tan\vartheta)}{\cos^2\beta \cdot (\tan\beta - \tan\alpha) \cdot (\tan\beta - \tan\vartheta)} \tag{5.14}$$

$$n_\vartheta = \frac{n_x \cdot \tan\alpha \cdot \tan\beta + n_y + n_{xy} \cdot (\tan\alpha + \tan\beta)}{\cos^2\vartheta \cdot (\tan\vartheta - \tan\alpha) \cdot (\tan\vartheta - \tan\beta)} \tag{5.15}$$

The choice of ϑ is restricted by the condition that a negative compression force n_ϑ has to result, as the concrete cannot sustain any tensile forces.

- If either n_α or n_β is negative (compressive force in the reinforcement) this force can be reduced to zero by varying ϑ (Fig. 5.6). At the same time the other reinforcement force is reduced.

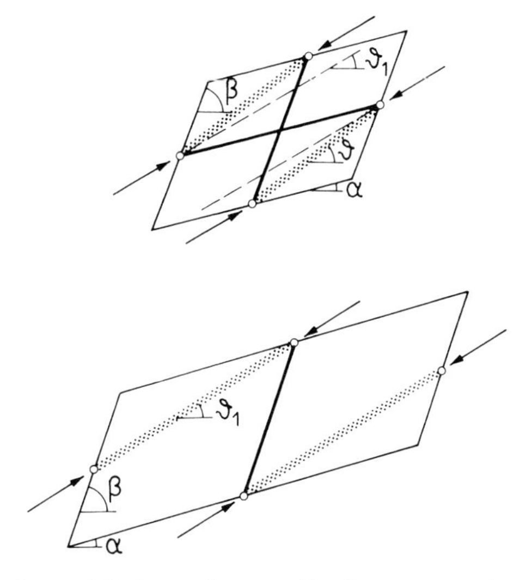

Fig. 5.6: Variation of ϑ in order to eliminate negative reinforcement force

- If n_α and n_β are positive a variation of ϑ causes a reduction of the force in one reinforcement and an increase of the force in the other. The same relationship holds for the web of a beam as shown in Fig. 2.27. The internal structural action can thus be selected according to an appropriate criterion (e.g. smallest amount of reinforcement, smallest intensity of concrete compression or behavior at working load).

- The required amounts of reinforcement can be obtained from the following conditions:

$$a_{s\alpha} \cdot f_y \geq n_\alpha \tag{5.16}$$

$$a_{s\beta} \cdot f_y \geq n_\beta \tag{5.17}$$

- The required thickness of the plate is given by:

$$t \cdot f_{ce} \geq -n_\vartheta \tag{5.18}$$

5.2 SLAB ELEMENTS

For design purposes a slab element can be subdivided into two plane stress elements as shown in Fig. 5.7.

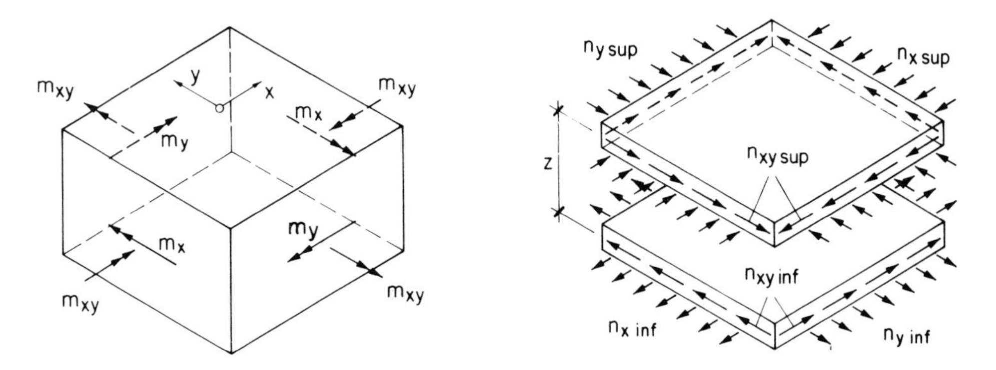

Fig. 5.7: Slab element subdivided into two plane stress elements

The stresses in the two plane stress elements result from the division of the bending moment by the internal lever arm z:

$$n_{x,inf} = m_x / z \qquad n_{x,sup} = -m_x / z \tag{5.19}$$

$$n_{y,inf} = m_y / z \qquad n_{y,sup} = -m_y / z \tag{5.20}$$

$$n_{xy,inf} = m_{xy,inf} / z \qquad n_{xy,sup} = -m_{xy} / z \tag{5.21}$$

The lever arm is usually estimated and then possibly improved iteratively by fully utilizing the compressive strength in the plane stress element. The design forces in the reinforcement and the concrete can be determined using the equations derived in the previous section. It should be observed that both compression field inclinations ϑ_{sup} and ϑ_{inf} can be chosen independently of each other.

A transverse shear force in the slab has to be resisted by a web element between the two plane stress elements. Since for slabs the shear force is in general small, no shear reinforcement is generally necessary, since the concrete tensile strength suffices to resist the corresponding tensile forces. However, if the shear force is large the web element between the plane stress elements can be designed like a beam web.

In Fig. 5.8 the stress resultants and the corresponding design forces acting on an element with an orthogonal reinforcement in the x and y directions are shown.

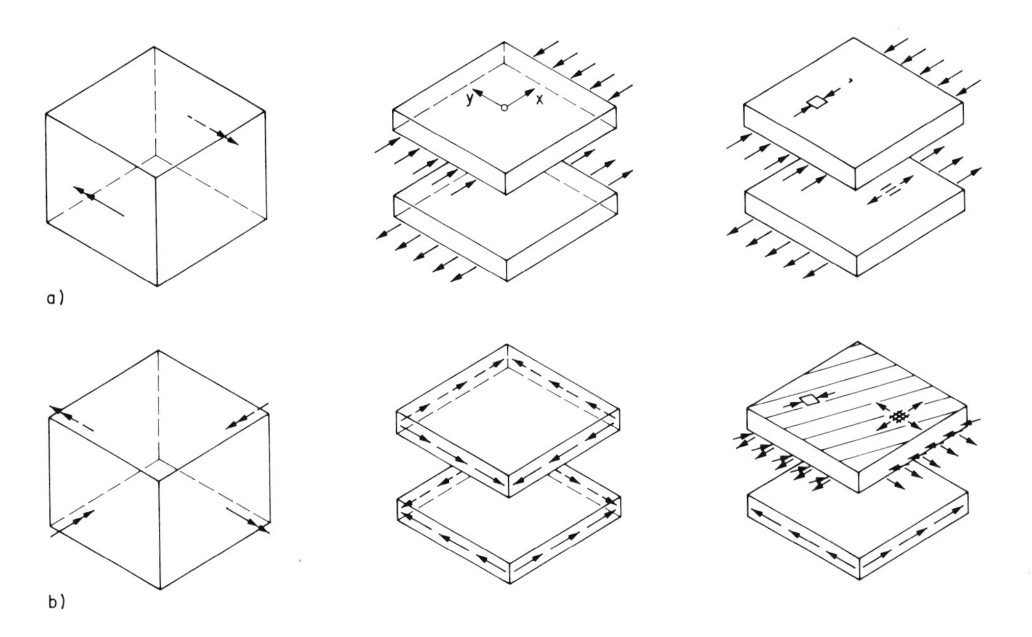

Fig. 5.8: Stress resultants and design forces in slab element with orthogonal reinforcement due to m_x, m_y and m_{xy}

These stress fields permit a visual consideration of the equilibrium conditions and the internal stresses at the edge of the slab. As shown in Fig. 5.9a a twisting moment m_{xy} is present in an edge region free of external torsional reactions, which is in equilibrium with the shear forces on the horizontal edges of an additional vertical element. This element resists the shear force along the edge of the slab and has to be provided, therefore, with vertical reinforcement. From the stress field it is evident that the horizontal reinforcement has to be fully anchored at the edge of the slab. These two conditions can be fulfilled by arranging the reinforcement in loops. The deviation forces for this reinforcement serve to change the strut action from the horizontal to the vertical element (Fig. 5.9b). In the corner region, where the two vertical elements touch, the vertical shear forces are in equilibrium with the support force (Fig. 5.9).

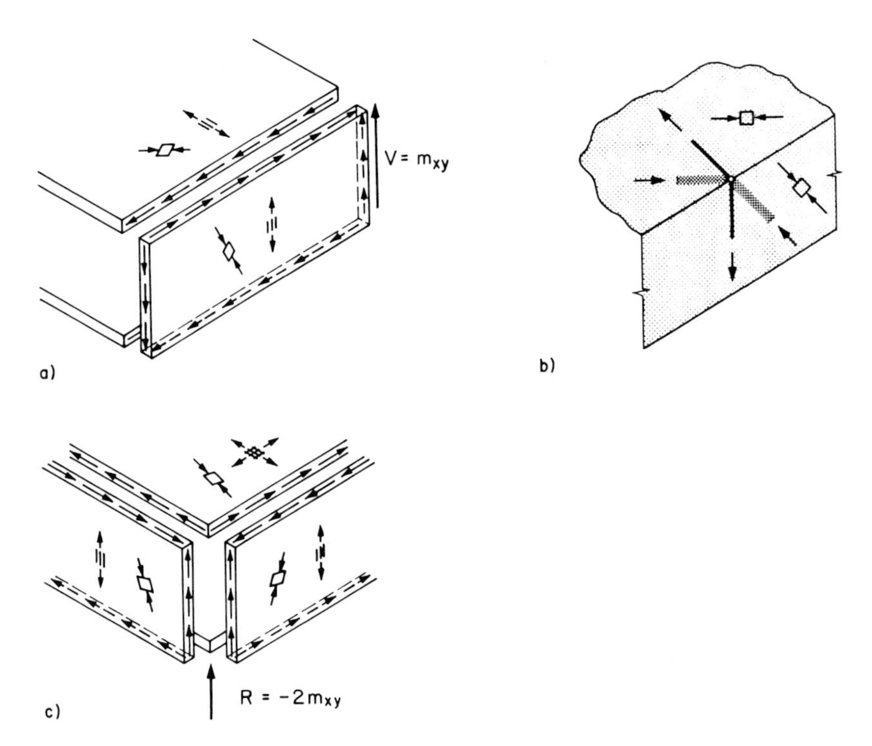

Fig 5.9: Stress situation at a free edge and corner of a slab caused by a twisting moment

5.3 SHELL ELEMENTS

In the case of shell elements, as for slab elements, the bending moments and membrane forces can be resisted by means of two plane stress elements (Fig. 5.10).

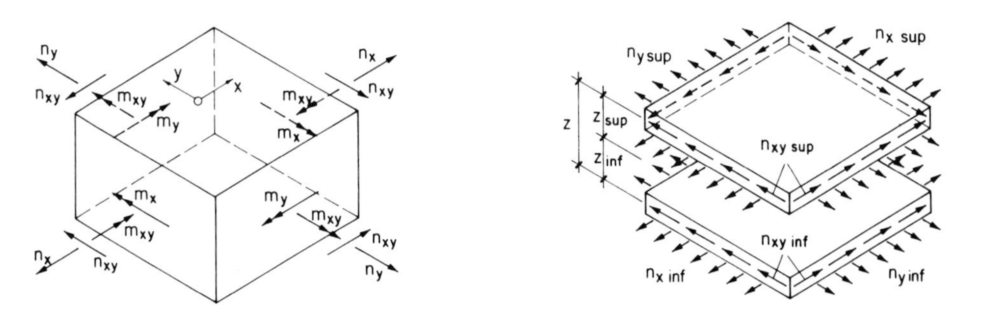

Fig. 5.10: Shell subdivided into two plane stress elements

The stresses in the two plane stress elements are determined as follows:

$$n_{x,inf} = (m_x + n_x \cdot z_{sup}) \;/\; z \qquad n_{x,sup} = (-m_x + n_x \cdot z_{inf}) \;/\; z \tag{5.22}$$

$$n_{y,inf} = (m_y + n_y \cdot z_{sup}) \;/\; z \qquad n_{y,sup} = (-m_y + n_y \cdot z_{inf}) \;/\; z \tag{5.23}$$

$$n_{xy,inf} = (m_{xy} + n_{xy} \cdot z_{sup}) \;/\; z \qquad n_{xy,sup} = (-m_{xy} + n_{xy} \cdot z_{inf}) \;/\; z \tag{5.24}$$

For the determination of the design forces the equations for the plane stress elements are again valid.

In Fig. 5.11, the example described in section 2.4.3 is shown, in which a shear loded web is subjected to a transverse bending moment. Due to the eccentricity of the displaced shear field the lateral bending moment can be resisted (see also [37)]. It should be observed that thereby a twisting moment results. But the latter is of little significance and simply causes a small shift of the line of action of the shear introduced into the web plate.

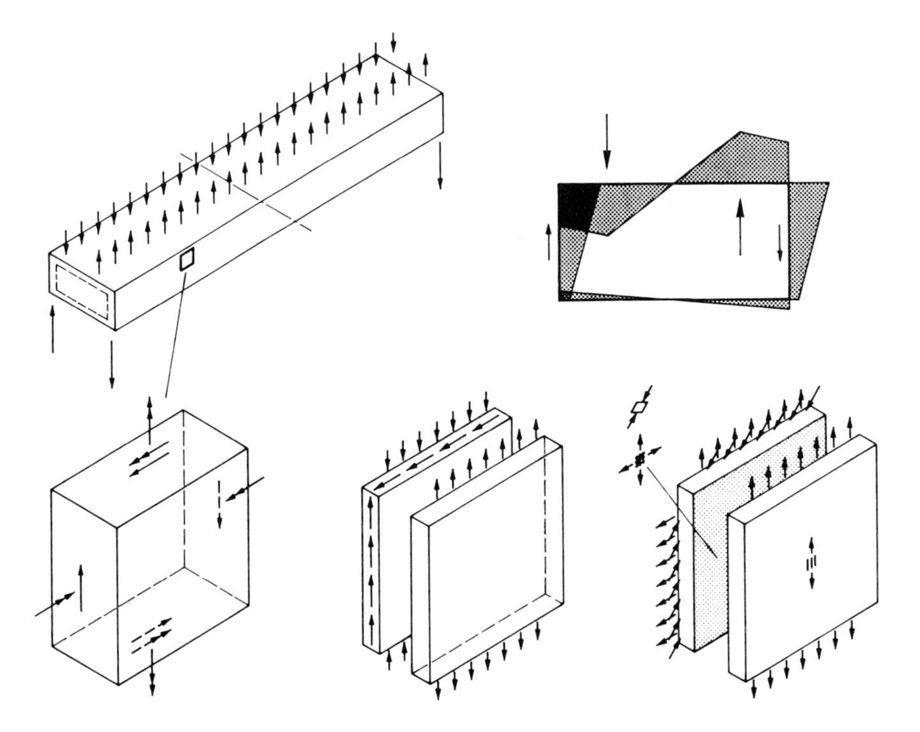

Fig. 5.11: Shear loaded web subjected to transverse bending

6 OUTLOOK: COMPUTER PROGRAMS

So far all numerical examples have been presented without any recourse to computer programs. In many, if not to say in most practical cases the use of hand calculations possibly supported by some auxiliary programs are completely sufficient and also economically competitive.

However it is recognized that computer programs to implement these methods should be developed. It is beyond the scope of this book to deal in detail with the problems involved in such an undertaking. However a few comments concerning the peculiarities of the mathematical problems involved as well as references to pertinent research studies may serve as a first orientation.

The highly developed elastic finite element analysis programs require the solution of a set of linear equations for which efficient methods are available. The mathematical formulation of the development of stress fields using the theory of plasticity leads to a linear program, for minimizing an objective linear function while satisfying both linear equations and inequality constraints. Its solution demands specially adapted methods to develop usable programs.

M. Schlaich [38] worked out an interactive computer program on the basis of discrete truss models. It offers the options of the elastic solution, the plastic solution, the shake-down load and an optimum solution, i.e. minimum reinforcement.

For the design of reinforced concrete walls, i.e. plates loaded in plane, R. Hajdin [39] uses triangular, uniformly stressed elements and linear reinforcing elements to create, interactively, statically admissible stress fields.

Recently Anderheggen and al. [40] presented a new approach starting with a linear elastic finite element solution. Sets of residual strains are then introduced to relax the highly stressed regions and to redistribute the internal forces in an appropriate manner. Z. Despot [41] wrote an interactive program for plates loaded in plane, whereas P. Steffen [42] developed a program for reinforced slabs loaded transversly. He also demonstrated that in most cases savings in reinforcement of up to one third can be achieved by using such a plastic redistribution of the internal forces. Both programs, despite the fact that they were developed for research use, have reached a remarkable state of user friendliness. However a basic knowledge of the theory of plasticity and the intuition and visualization of stress field solutions, gained by the study of the methods presented in this book, and their practical application using hand calculations are most useful for a proper interpretation of the computer results.

The authors are very pleased that a promising start has been made to provide practical computer programs to implement the design of reinforced concrete structures using stress fields.

REFERENCES

[1] Mac Gregor J.G.: *Challenges and Changes in the Design of Concrete Structures*, Concrete International, 1984.

[2] Melan E.: *Der Spannungszustand eines Mises-Henckyschen Kontinuums bei veränderlicher Belastung*, Sitz. ber. Akad. Wiss. Wien, Abt. IIa, 147, 1938.

[3] Maillart R.: *Aktuelle Fragen des Eisenbetonbaues*, Schweizerische Bauzeitung 56, 1938.

[4] Mörsch E.: *Der Eisenbetonbau - Seine Theorie und Anwendung*, Verlag von Konrad Wittwer, Stuttgart, 3. Aufl. 1908, 4. Aufl. 1912, 5. Aufl. 1.Bd., 1.Hälfte, 1920, 2. Hälfte 1922.

[5] Prager W., Hodge P.G.: *Theory of Perfectly Plastic Solids*, John Wiley & Sons, Inc., New York, 1951.

[6] Ritter W.: *Die Bauweise Hennebique*, Schweizerische Bauzeitung 17, 1899.

[7] Drucker D.C.: *On Structural Concrete and the Theorems of Limit Analysis*, International Association for Bridge and Structural Engineering (IABSE), Zürich, Abhandlungen 21, 1961.

[8] Thürlimann B., Grob J., Lüchinger P.: *Torsion, Biegung und Schub in Stahlbetonträgern*, Institute of Structural Engineering, ETH Zürich, Autographie zum Fortbildungskurs für Bauingenieure, 1975.

[9] Thürlimann B., Marti P., Pralong J., Ritz P., Zimmerli B.: *Anwendung der Plastizitätstheorie auf Stahlbeton*, Institute of Structural Engineering, ETH Zürich, Autographie zum Fortbildungskurs für Bauingenieure, 1983.

[10] Nielsen M.P.: *Limit Analysis and Concrete Plasticity*, Prentice-Hall, Englewood Cliffs, New Jersey, 1984.

[11] Chen W.F.: *Limit Analysis and Soil Plasticity*, Elsevier Scientific Publishing Company, Amsterdam, Oxford, New York, 1975.

[12] Müller P.: *Plastische Berechnung von Stahlbetonscheiben und -balken*, Institute of Structural Engineering, ETH Zürich, Report No. 83, Birkhäuser Basel, Stuttgart, 1978.

[13] Marti P.: *Basic Tools of Reinforced Concrete Beam Design*, Journal of the American Concrete Institute, Vol. 82, No. 1, Jan.-Feb. 1985. Discussion, Vol. 82, No. 6, Nov.-Dec. 1985.

[14] Collins M.P., Mitchell D.: *Shear and Torsion Design of Prestressed and Non-Prestressed Concrete Beams*, Journal of the Prestressed Concrete Institute, Vol. 25, No. 5, Sept.-Oct. 1980. Discussion, Vol. 26, No. 6, Nov.-Dec. 1981.

[15] Schlaich J., Weischede D.: *Ein praktisches Verfahren zum methodischen Bemessen und Konstruieren im Stahlbetonbau*, Comité Euro-International du Béton (CEB), Bulletin d'Information No. 150, Paris, 1982.

[16] Schlaich J., Schäfer K. und Jennewein M.: *Toward a Consistent Design of Structural Concrete*, Journal of the Prestressed Concrete Institute, Vol. 32, No. 3, 1987.

[17] Exner H.: *On the Effectiveness Factor in Plastic Analysis of Concrete*, Internationale Vereinigung für Brückenbau und Hochbau (IVBH), Schlussbericht Kolloquium Kopenhagen, Zürich, 1979.

[18] Menne B.: *Zur Traglast der ausmittig gedrückten Stahlbetonstütze mit Umschnürungsbewehrung*, Deutscher Ausschuss für Stahlbeton, Heft 285, Wilhelm Ernst & Sohn, Berlin, 1977.

[19] Muttoni A.: *Die Anwendbarkeit der Plastizitätstheorie in der Bemessung von Stahlbeton*, Institute of Structural Engineering, ETH Zürich, Birkhäuser Basel, Boston, Berlin, 1990.

[20] Ferguson P.M., Thompson J.N.: *Development Length of High Strength Reinforcing Bars in Bond*, Journal of the American Concrete Institute, Proc. V. 59, 1962.

[21] Ferguson P.M., Thompson J.N.: *Development Length for Large High Strength Reinforcing Bars*, Journal of the American Concrete Institute, Proc. V. 62, 1965.

[22] Chamberlin S.J.: *Spacing of Reinforcement in Beams*, Journal of the American Concrete Institute, Proc. V. 53, 1956.

[23] Robinson J.R., Zsutty T.C., Guiorgadze G., Lima L.J., Jung H.L., Villatoux J.P.: *La Couture des Junctions par Adhérence*, Annales de l'Institut Technique du Bâtiment et des Travaux Publics, Supplément du No. 318, Paris, 1984.

[24] Chinn J., Ferguson P.M., Thompson J.N.: *Lapped Splices in Reinforced Concrete Beams*, Journal of the American Concrete Institute, Proc. V. 52, 1955.

[25] Ferguson P.M., Breen J.E.: *Lapped Splices for High Strength Reinforcing Bars*, Journal of the American Concrete Institute, Proc. V. 62, 1965.

[26] Chamberlin S.J.: *Spacing of Spliced Bars in Beams*, Journal of the American Concrete Institute, Proc. V. 54, 1958.

[27] Ferguson P.M., Briceno E.A.: *Tensile Lap Splices - Part 1: Retaining Wall Type, Varying Moment Zone*, Research Report 113-2, Center for Highway Research, The University of Texas at Austin, 1969.

[28] Ferguson P.M., Krishnaswamy C.N.: *Tensile Lap Splices - Part 2: Design Recommendations for Retaining Wall Splices and Large Bar Splices*, Research Report 113-3, Center for Highway Research, The University of Texas at Austin, 1971.

[29] Thompson M.A., Jirsa J.O., Breen J.E., Meinheit D.F.: *The Behavior of Multiple Lap Splices in Wide Sections*, Research Report 154-1, Center for Highway Research, The University of Texas at Austin, 1975.

[30] Tepfers R.: *A Theory of Bond Applied to Overlapped Tensile Reinforcement Splices for Deformed Bars*, Publication 73.2, Division of Concrete Structures, Chalmers University of Technology, Göteborg, 1973.

[31] Rathkjen A.: *Forankringsstyrke af armeringsjern ved bjaekeunderstøtninger*, Rapport 7203, DIAB Aalborg, Ren & Anvendt Mekanik, 1972.

[32] Hess U.: *The Anchorage of Reinforcing Bars at Supports*, Internationale Vereinigung für Brückenbau und Hochbau, Schlussbericht Colloquium Kopenhagen, Zürich, 1979.

[33] Stucki D., Thürlimann B.: *Versuche an Eckverbindungen aus Stahlbeton*, Institute of Structural Engineering, ETH Zürich, Report No. 8701-1, Birkhäuser Basel, Boston, Berlin, 1990.

[34] Hillerborg A.: *Strip Method of Design*, Cement and Concrete Association, Viewpoint Publication 12.067, Wexham Springs, 1975.

[35] Marti P.: *Gleichgewichtslösungen für Flachdecken*, Institute of Structural Engineering, ETH Zürich, Report No. 117, Birkhäuser Basel, Boston, Stuttgart, 1981.

[36] Muttoni A., Schwartz J., Thürlimann B.: *Bemessen und Konstruieren von Stahlbetontragwerken mit Spannungsfeldern*, Institute of Structural Engineering, ETH Zürich, Lecture Notes, 1987.

[37] Thürlimann B.: *Schubbemessung bei Querbiegung*, Schweizerische Bauzeitung, Heft 26, 1977.

[38] Schlaich M.: *Computerunterstützte Bemessung von Stahlbetonscheiben mit Fachwerkmodellen*, Report No. 1, Institute of Informatic, ETH Zürich, 1989.

[39] Hajdin R.: *Computerunterstützte Berechnung von Stahlbetonscheiben mit Spannungsfeldern*, Institute of Structural Engineering, ETH Zürich, Report No. 175, Birkhäuser Basel, Boston, Stuttgart, 1990.

[40] Anderheggen E., Despot Z., Steffen P. N., Tabatabai S. M. R.: *Finite Elements and Plastic Theory: Integration in Optimum Reinforcement Design*, Proceedings of the sixth international conference on computing in civil and building engineering, Berlin, Germany, July 1995, A. A. Balkema, Rotterdam, Brookfield, 1995.

[41] Despot Z.: *Methode der Finiten Elemente und Plastizitätstheorie zur Bemessung von Stahlbetonscheiben*, Institute of Structural Engineering, ETH Zürich, Report No. 215, Birkhäuser Basel, Boston, Stuttgart, 1995.

[42] Steffen P.: *Elastoplastische Dimensionierung mittels Finiter Bemessungselemente und linearer Optimierung*, ETH Zürich, Ph.D. thesis No. 11611, 1996.

INDEX